# The Tsunami of 2004 in Sri Lanka

This book is based on empirical research in Sri Lanka conducted after the catastrophic tsunami which hit the country in December 2004. The aims of the research have been to develop new knowledge on post-crisis reconstruction and recovery work, on how to bridge the knowledge gap between researchers and practitioners, as well as trying to use past research experiences from Sri Lanka to learn about the present day situation. The chapters use a common analytical frame related to the 'policy narratives' of post-tsunami recovery in the shadow of war, and deal with housing reconstruction, livelihoods, internally displaced, humanitarian interventions and protracted conflicts.

The authors represent various social scientific fields and they have experience from different geographical areas of Sri Lanka.

This book was published as a special issue of *Norsk Geografisk Tidsskrift*.

**Piers M. Blaikie** is Emeritus Professor in the School of International Development at the University of East Anglia, Norwich, United Kingdom.

**Ragnhild Lund** is a Professor in the Department of Geography at the Norwegian University of Technology, Trondheim, Norway.

# The Tsunami of 2004 in Sri Lanka

## Impacts and Policy in the Shadow of Civil War

Edited by Piers M. Blaikie and Ragnhild Lund

Routledge
Taylor & Francis Group
LONDON AND NEW YORK

First published 2010 by Routledge
2 Park Square, Milton Park, Abingdon, Oxon, OX14 4RN

Simultaneously published in the USA and Canada
by Routledge
270 Madison Avenue, New York, NY 10016

*Routledge is an imprint of the Taylor & Francis Group, an informa business*



Typeset in Times by Value Chain, India
Printed and bound in Great Britain by TJ: Digital Ltd, Padstow, Cornwall



*British Library Cataloguing in Publication Data*
A catalogue record for this book is available from the British Library

ISBN10: 0-415-55909-X
ISBN13: 978-0-415-55909-6

# CONTENTS

# Editorial

# The tsunami of 2004 in Sri Lanka: Impacts and policy in the shadow of civil war

During the past decade the *NTNU Research Group on Forced Migration*, which is located at the Department of Geography, Norwegian University of Science and Technology (NTNU), has worked with several research projects in Sri Lanka. Much of the work has been carried out after the catastrophic tsunami hit the country in December 2004. The group has worked with NGOs on various aspects of capacity building, with local research partners from the University of Peradeniya, the South Eastern University of Sri Lanka, and the Eastern University, Sri Lanka, and also researchers from various disciplines at NTNU. The research aims have been to develop new knowledge on what constitutes post-crisis reconstruction and recovery work, on how to bridge the knowledge gap between researchers and practitioners, as well as trying to use past research experiences from Sri Lanka to learn about the present day situation. This book is one outcome of these activities, and represents an effort to synergise and build upon the knowledge and perspectives of researchers directly linked to the research activities of the forced migration group and associated partners. Encouraged by Piers Blaikie's idea to work on various policy narratives of post-tsunami recovery in the shadow of war (see chapter one), the authors have been inspired to analyse policy processes and policy narratives related to their field of expertise and the various facets of reconstruction and recovery work after the tsunami.

Within the broad brief of the Research group on Forced Migration, the two intertwined disasters of tsunami and civil war in Sri Lanka form the focus of this book. As Chapter One will explain in more detail, it may seem that a tsunami (an event taking place over a few minutes) and a protracted civil war over more than twenty years could not be more different in terms of impact and policy. However, they are linked in a number of inseparable ways. Firstly, in both cases there is a long history of the reproduction of inequality and vulnerability, which renders certain people (differentiated by age, sex, kinship, ethnic identity, livelihood, wealth, political networks) more vulnerable to disaster than others. Of course it is obvious that a civil war and a tsunami will have quite different impacts but the underlying patterns of vulnerability on which these two different disasters develop have similar histories of ethnic tension, discrimination and dispossession. Secondly, the impacts of the war and the tsunami are inextricably linked. Patterns of mortality, loss of home and livelihood and displacement following the tsunami are not formed on *a tabula rasa but overlay* those caused by the war. Therefore, in many cases and locations (particularly in the north and east of Sri Lanka and in coastal regions) many – but not all – the impacts of the war and the tsunami have to be understood in the context of each other. Thirdly, many of the policies which relate to the effects of the war and the tsunami also have common origins. After all the Government of Sri Lanka is the final arbiter of policies which address both issues and many of the international, bi-lateral and local actors which have some influence on policy are the same.

However, while we were working on this project, the relative importance of these two disasters altered. The war in Sri Lanka intensified, and we experienced a situation in which the issue of the tsunami was almost forgotten in a maze of new policies which were shaped during the final stage of the brutal war in the LTTE (Tamil Tigers Liberation Movement) controlled areas. Sri Lanka's government on May 18 declared victory over the LTTE and later the said founder Vellupalai Prabhakaran was killed. That effectively ended the LTTE's struggle to create a separate state for minority Tamils, which began in the 1970s and erupted into civil war in 1983. What the war has meant for the post-tsunami recovery work going on in the country still remains to be seen.

As of the date of writing this Introduction (July 2009), debates on the new challenges related to the post-war reconstruction dominate the media. The post-tsunami recovery work has come to a halt and the voices of those who have still not been compensated for their losses during and after the tsunami have been muted. This is also a tragedy, as a substantial number of people who have been displaced after the tsunami have not yet been given compensation and houses, their livelihoods have not been restored, and individuals and whole communities are still traumatised. Bearing this in mind, what constitute the major challenges in post-war Sri Lanka?

One major challenge is to handle the humanitarian crisis which prevails in the north of the country. Throughout the war, and particularly during its final stage, civilians displaced by the conflict were facing a severe crisis. Hundreds of thousands of people were displaced from the north within a so-called 'No-fire Zone' with extremely limited humanitarian relief. The Sri Lankan government posed severe restrictions on international humanitarian organisations in the conflict zone in September 2008, so very limited humanitarian space limited their access to the displaced (iDMC 2009). At the end of the war the human costs were huge. According to recent, albeit uncertain estimates, nearly 24,000 soldiers were killed and 5,000 left permanently disabled (AlertNet1 2009); between 80,000-100,000 people were believed killed during the war, and exceeding 20,000 during the final month of the government's siege against the LTTE (AlerntNet2 2009); and about 290,000 Tamil civilians remain inside camps (AlertNet3 2009).

Al Jazeera (2009) reported on July 6 2009 that the latter stages of the war in Sri Lanka have been '*carefully choreographed and hidden from the outside world, with the voices of the victims silenced through fear and insecurity'*. Aid workers have for some time now called for unhindered access to more than 40 camps for internally displaced persons in Sri Lanka's north. In spite of urgent demands on a very large scale for medical care, food, water and other essentials, access to these camps still remain restricted. Aid workers who have been permitted access say that minimum humanitarian standards are not being met. One of the reasons is that the government is trying to weed out any possible LTTE fighters and sympathisers, and daily reports about suspected former Tamil rebel fighters are being sent for 'rehabilitation and vocation training'. The press as well as most humanitarian and donor agencies are still being prevented from getting access to these areas. This state of affairs stands in sharp contrast to President Rajapakse's 'victory speech' that told Sri Lanka's parliament that 'our heroic forces have sacrificed the lives to protect Tamil civilians' and that he would take 'personal responsibility' for protecting Tamils. In spite of such statements, he is now scandalously confining this population in squalid and dangerously unhealthy conditions. According to Martin Shaw (see also The Economist 2009) the government has officially backtracked, and in spite of massive international pressure, on plans to hold the displaced while screening them for potential terrorists, albeit the government now says that 80 percent will be resettled by the end of 2009. However, Shaw (2009,1) in citing Human Rights Watch comments on the new policy and control of internally displaced: *'The government's history of restricting the rights of displaced persons through rigid pass systems and strict restrictions on leaving the camps heightens concerns that they will be confined in camps much longer, possibly for years.'* Sadly, one is therefore not told the whole story about the humanitarian situation in the north and the situation in the 'concentration camps' (Roy 2009). Needless to say, the war-ravaged north needs massive reconstruction and recovery.

The major media debates as of July 2009 are e. about how to solve inequality and achieve progress in the country. President Rajapakse is in fact faced with a dual challenge: that of pleasing the Sinhalese majority of the population and win the upcoming local elections in Vavuniya and Jaffna (8 August 2009), and that of reducing social and economic inequality that fuelled the war and other insurgencies in the first place. According to Reuters (AlertNet4 2009), the effective end of the Tamil Tigers' rebellion in May 2009 is no guarantee of long-term peace unless the government can also deal with the common root causes that link the war with the LTTE with previous bloody Marxist uprisings. First, the Marxist (JVP) party, which was very much an influential partner in the war against LTTE, expects to be heard. It represents working-class Sinhalese from the rural south who fought the government both in the 1970s and late 1980s/early 1990s, and drew popular support from a platform of economic equality. Actually, there was about as many as were killed during JVP insurrections as during the 25 year LTTE war. Secondly, the situation in eastern Sri Lanka is turbulent as well. Being the most ethnically mixed area of the nation where Sinhalese, Muslims and Tamils are all strongly represented, poverty still persists and various youth groups are presently forming militias. Land that was evacuated because of the tsunami is now being expropriated for the purpose of constructing a harbour, tourist resorts and industrial activities and thus erodes the traditional fishing activities. This development puts long term social stability at risk and the vulnerability of the present situation into perspective. As says Reuters (AlertNet1 2009,1):

*'A boom in the Colombo Stock Exchange linked to government success in the war, falling interest rates and anticipated foreign direct investments are cold comfort to many of Sri Lanka's 21 million people. Most depend on agriculture* [and fishing] *and want post-war progress faster than trickle down effects from economic liberalisation are likely to deliver...Census and Statistics Department data show that 40 percent of Sri Lanka's people have only 13 percent of total income, while the richest fifth enjoy 54 percent.'*

So far, there is still aversion in the country against international (generally western) donors and humanitarian agencies to facilitate the reconstruction and recovery processes. However, a new picture of 'donors' emerge from the Asian region itself, such as the Indian state and commercial Chinese enterprises. The Indian government, for example, has set aside Rs 500 crore for rehabilitation of the internationally displaced people and reconstruction of areas in Northern and Eastern Sri Lanka devastated by the war against the LTTE. Sri Lanka has also granted China an exclusive economic zone in an effort to attract more investments from the Asian wealthy nations (Lanka Academic Network 2009). The Chinese are also funding a port and a coal power plant in the south, and the President is pushing for change in agriculture by bringing in the expertise of M. S. Swaminathan (the father of the Green Revolution) to boost local production and cut dependence on imports (AlertNet4 2009, Lanka Academic Network 2009).

However, the country still seems to be in a state of flux and vulnerable to future shocks. This brings us back full circle to where we started our work, in trying to understand impacts and policies of two contingent disasters (tsunami and war). With respect to the new post-war challenges, we believe that our knowledge may contribute valuably to the understanding of how recovery works, and the policy processes and policy narratives involved. New post-war narratives may arise and replace the old ones that we have dealt with, while others will endure. However, even more so, it may be found that the new narratives bear many resemblances to the post-tsunami ones, and as such the book provides valuable insights and lessons learnt from recovery work in general.

The story continues to unfold as we speak, and at a pace which a book is not able to capture. We authors have to draw a line in time in Sri Lankan history where the future is particularly uncertain. Will positive development take place to address the long running grievances and ethnic antagonisms that lie at the root of the old conflict, or, will the present parliamentary leaders lose this opportunity for the sake of political power and ethnic dominance?

## References:

Al Jazeera 2009 'No welfare for Sri Lanka's Tamils' in *Lanka Academic Network*, July 6 2009.

AlerNet 1 2009 http://www.alertnet.org/thenews/newsdesk/COL473976.htm?refer=wd20090603, accessed 2009-07-06.

AlerNet 2 2009 http://www.alertnet.org/thenews/newsdesk/NO1355468.htm?refer=wd20090603, accessed 2009-07-06.

AlertNet 3 2009 http://www.alertnet.org/thenews/newsdesk/IRIN/36fde245298876e92bc4ae7670c61cf..., accessed 2009-07-06.

AlertNet4 2009 http://www.alertnet.org/thenews/newsdesk/COL423696.htm, accessed 2009-07-06.

*iDMC* (internal displacement monitoring centre) 2009 http://www.internal-displacement.org/8025708F004CE90B/(httpCountrySummaries)/..., accessed 1 May 2009.

*Lanka Academic Network* 2009 Monday July 6 07:46:53 EDT 2009

*The Economist* 2009 'Tainted triumph', 23 May 2009

Roy, Aninditha 2009 'Sri Lanka's genocidal warfare' *The Guardian Weekly* 10 April 2009.

Shaw, Martin 2009 http://www.opendemocracy.net/article/sri-lanka-camos-media-genocide, accessed 30 June 2009.

*Professor Ragnhild Lund*
*Project coordinator*
*NTNU Research Group on Forced Migration*
*Department of Geography*
*Norwegian University of Science and Technology*
*NO-7491 Trondheim, Norway*
*E-mail: ragnhild.lund@svt.ntnu.no*

INTRODUCTION

# The tsunami of 2004 in Sri Lanka: An introduction to impacts and policy in the shadow of civil war

PIERS M. BLAIKIE

The tsunami of 2004 and ongoing civil war in Sri Lanka are treated in the Special Issue of *Norsk Geografisk Tidsskrift–Norwegian Journal of Geography* as two inter-related disasters. Both have been shaped by long-established patterns of vulnerability. Understanding their outcomes and offering constructive critique about policy responses depend both on explanations and models of social processes as well as an understanding of the post-tsunami policy process in the shadow of civil war. A framework is offered for analysing the policy process to which the articles in the Issue contribute. Policy process is examined in terms of policy narratives, the actors involved and their institutions. Two policy narratives constructed by international development institutions – vulnerability, and livelihood destruction and recovery – are discussed, followed by more focused policy narratives concerning internally displaced persons (IDPs), housing policy, resettlement, and the 'right to return'. They are reproduced dynamically between a range of actors, including those in civil society (differentiated by ethnicity, gender, age, wealth, social and political capital, and geographical location), international non-governmental organisations (INGOs), non-governmental organisations (NGOs), community-based organisations (CBOs), civilian government, the military, and others. Policy narratives evolve and interact reflexively with the policy process and address the impact and longer term development effects of the tsunami and the civil conflict that overshadows it.

*Piers M. Blaikie, Emeritus Professor, School of Development Studies, University of East Anglia*

## Two disasters

This Special Issue of *Norsk Geografisk Tidsskrift–Norwegian Journal of Geography* focuses on the impacts of the tsunami of 26 December 2004 in Sri Lanka and associated policy issues of humanitarian assistance and recovery. However, this disaster unfolded in the shadow of a much longer one, the civil war, which has been waged, except at times of ceasefire, since *c.*1983. It is impossible to understand the impacts of the tsunami and the policy response without putting it into the overall and all-too-pervasive context of the civil war. These two disasters clearly continue to interact in terms of socio-economic outcomes and of policy. The humanitarian response to both the tsunami and the war has been and continues to be shaped by the Government of Sri Lanka (GoSL) in negotiation with international and bilateral donors, and constitutes one of the main interfaces at which policy negotiations take place.

The tsunami of 24 December 2004 killed about 36,000 people in Sri Lanka and injured 21,000 in a matter of minutes and, in the longer term, displaced between 450,000 and 516,000, and destroyed or damaged nearly 100,000 houses and 200 educational facilities (GoSL 2005). The civil war over a period of 25 years has caused the deaths of 70,000 people. It has generated between 500,000 and one million internally displaced persons of the total population of *c.*20 million and led to the emigration of an estimated 800,000 (Schrijvers 1999). The disasters of the tsunami and civil war, although very different in their genesis and outcomes, also have similarities. They are both slow in unfolding. Even the tsunami, a natural event of a few minutes, became a disaster only as a result of the emergence and reproduction of vulnerability over a long period before its initial devastating impact (Kelman & Gaillard 2007; Khasalamwa 2009, this Issue). Vulnerability is a much debated term and is reviewed by Khasalamwa (2009) and John Nigel (2009, this Issue). A simple but widely accepted version of a definition is:

> Vulnerability is a set of characteristics of a person or group of persons in terms of their capacity to anticipate, cope with, resist and recover from the impact of natural hazards. (Blaikie et al. 1994)

The term can be reworked to relate to a war where the impact is usually more spatially widespread, more long lasting and multivalent than a single environmental event such as an earthquake, flood or a tsunami. Therefore, in a long war, as in Sri Lanka, people of different ethnicities, wealth, age, and gender are constantly trying to anticipate, cope with, resist, and recover from the impact of an ebbing and flowing war. Vulnerability emerges through a number of socio-economic and political forces and often pushes economically and socially marginal people into marginal places, which may be hazardous in terms of environmental conditions and discrete events. Thus, the war and the tsunami as the 'threat to come' have intertwined social and political antecedents. Khasalamwa (2009) shows how the war has created vulnerabilities through death and disappearance of breadwinners, displacement of agricultural activities, destruction of industrial plant, and thereby of employment opportunities. All this in turn has rendered those in the coastal zone especially vulnerable to the effects of the tsunami. Also, there has been a tendency for those with less access to livelihood opportunities and political support during hostilities (typically some Muslims and Tamils) to be displaced towards the poorer agricultural land near the coast (Hyndman 2009, this Issue; Khasalamwa 2009) Further, the buffer zone declared by the GoSL disallowed rebuilding of houses within a 200 m zone along the east coast and discriminated heavily against Tamils and Muslims who had fled the war and settled in the less fertile and less sought-after land close to the coast, which then bore the brunt of the tsunami (Hyndman 2009).

In more general economic and social terms, John Nigel (2009) highlights the overriding importance of ethnicity in the creation and reproduction of vulnerability in the conflict environment by comparing patterns of vulnerabilities in three villages of different ethnicities.

These patterns of vulnerability to the threats of war and the tsunami have emerged over a matter of generations although, in the case of the tsunami, the natural event ('the trigger') may have been a matter of minutes or days. Thus, the tsunami disaster is social in its genesis and particularly in its aftermath in terms of humanitarian aid and reconstruction, as Hewitt (1983), Blaikie et al. (1994), Wisner et al. (2004), and others have long pointed out in generic terms, and also Khasalamwa (2009) specifically in relation to Sri Lanka. The war through military action renders people vulnerable to attack, death, wounding, loss of livelihood, and loss of security (even to till fields and fetch essential supplies), loss of networks of support and also medical and educational services, all of which in turn reproduce vulnerability. The production of vulnerability in this war is vividly and insightfully discussed by Lang & Knudsen (2009, this Issue); Tete (2009, this Issue) and John Nigel (2009). The war constantly involves ceasefires, resumptions of hostilities, and new military strategies, and these changing threats are faced by a differentiated range of people with vulnerabilities reproduced in the past – and in the case of war in the immediate past, as hostilities and specific military action take new forms. There is the special case of the vulnerabilities of internally displaced persons (IDPs) through war (and internal displacement can happen a number of times over the period of conflict) and through the impact of the tsunami. Such displacements are driven by individual, family and community decisions in the face of particularly threatening situations and by resettlement policy shaped by government, and the latter is usually heavily freighted with political and strategic considerations (discussed by Tete 2009 and Hyndman 2009). The unresolved political nature of the war continues to reproduce vulnerabilities of (IDPs) despite, and sometimes because of, return to original areas where political and military contours have changed. In eastern Sri Lanka, civilians displaced by war often remain in areas of armed conflict and under threat, unable to escape politicisation, and are targeted on that basis. Lang & Knudsen (2009) have examined the protracted condition of insecurity for the internally displaced in the context of a live conflict.

## Analysing the policy process

Policy may briefly be defined as a set of stated intentions and resultant practices in the name of the public good. The policy process is the means by which policy is conceived, negotiated, expressed, and perhaps brought into law, and the procedures of implementation and practice. In the recovery process from the tsunami, policy has involved such issues as land use planning of the coastal zones (buffer zones), resettlement of IDPs, restoration of land rights, housing policy, income generation, micro-credit, refinancing the fishing sector, social services, and humanitarian protection for affected civilians. Many of these policies are discussed in this Issue. Understanding the policy process is essential for explaining outcomes in order to understand how the policy process actually (and often informally) takes place. Policy reform does not emerge as a linear response to 'truth talking to power' – as a result of facts from whatever source, including research, which reveal new truths and support alternative rational arguments for a policy change. The policy process is much more complex than this over-simplifying model, but which in a rational and positivist sense can be both powerful and useful. Policies are negotiated between a range of actors, in this case, different functionaries of the GoSL, the military, and other outside bodies (the organisations of the United Nations, INGOs (international non-governmental organisations), national NGOs (non-governmental organisations), and CBOs (community-based organisations). In most cases, it is not the policy itself which is the focus of negotiation but the broader political issues which the policy framework implicates. Silva (2009, this Issue) provides a penetrating analysis of the importance of the overall political environment in which a disaster unfolds by means of a comparison of the impact of a disaster in the 1930s (the malaria epidemic) and the tsunami of 26 December 2004. The former brought about a progressive political shift against British imperial rule and the beginnings of an impressive welfare state in what was to become independent Sri Lanka. The latter has not led to any progressive political changes and has merely underlined the relative failure of a much better funded international humanitarian response that has failed to galvanise the peace process. Specifically, the Post-Tsunami Operational Management Structure was proposed in 2005, but was undermined by the resistance of nationalist elements in the GoSL to the representation of the LTTE (Liberation Tigers of Tamil Elam) in the management structure, which was claimed to lend legitimacy to the LTTE. Other organisations replaced the Structure: first, the Task Force to rebuild the nation (TAFREN) and then the Reconstruction and Development Agency (RADA). Here, issues of political and military advantage as interpreted by the two main protagonists in the civil war met a wide range of ideological, moral and humanitarian issues from outside organisations. Negotiations between them and the GoSL and the subsequent protracted and complex outcomes in a fast-moving situation shaped policy, and later outcomes unfolded slowly in ways which may or may not bear any resemblance to agreed policy statements. Thus, the traditional (and sensible) question 'How can research be transferred to the policy sphere?' may be complemented by the question 'Why are some of the ideas that circulate in research and policy networks picked up and acted upon, while others are ignored and disappear?'

Therefore, a framework for understanding policy is necessary, and preferably one which is not totally unacceptable for a wide range of actors who are directly and indirectly involved in policy-making and implementation, such as academics, consultants and policy-makers and implementers at the local level. This is a tall order and one which may be impossible to satisfy. The framework presented here has two major elements – actors and their narratives. At all stages of the policy process, and particularly in implementation, the politics of knowledge production (here, especially development knowledge about the two disasters discussed in this Issue) is an important element throughout. It concerns the production of 'authoritative' knowledge about such issues as explanations of the past and possible future impacts of the tsunami, the statistics for what constitute displaced persons, priorities for immediate humanitarian aid, reconstruction, the 'peace dividend', sustainable livelihoods, IDPs, their rights, land-use zoning, etc. For example, a perception that humanitarian protection is the preserve of protection-mandated agencies works to prevent other agencies from embracing humanitarian protection as an approach to humanitarian assistance (Lang & Knudsen 2009). Another example from Lang & Knudsen (2009) is the emphasis of policy narratives on the duration of displacement rather than on the absence or failure of solutions. This hampers an understanding of displacement as tied to vulnerabilities and threats that endure in the context of protracted conflict. How knowledge is produced and communicated to others takes the discussion from the structural and institutional aspects of policy to the more post-structural, discursive and more overtly political aspects. For example, the way in which statistics about the tsunami are collected categorise and focus on particular ways of 'understanding the problem' (see Hyndman 2009, who examines the production and contestation of scientific knowledge regarding public safety and the changing regulations of the buffer zone in Sri Lanka). It is important, therefore, to examine in more detail the substance of how development knowledge and 'facts' are produced and accorded authority. Certain problems are framed and attention is invited to interact with propositions, facts and assumptions within the frame of this problem definition, but diverted from others which are then ignored and left out of the frame. There are key strings of propositions made by different actors in the policy process by way of policy argumentation and calls for examination of 'policy narratives' or stories told by different protagonists. These are not 'just talk' or inventions for others' amusement, but persuasive constructions with a beginning (assumptions, problem framing, choice of issues, etc.), a development (argumentation, supporting evidence, justifications, troublesome side issues and other relevant circumstances) and a conclusion (what should be done and policy recommendations). They use some facts, are ignorant of or deselect others and interpret information in a particular manner so as to tell a persuasive and consistent story. They frame issues and problems in certain ways to focus on some issues and to exclude others. This may be done either consciously as a strategy, or unconsciously, where a particular set of facts and values is not critically reflected upon, or, as is arguably more common, some of both. Narratives are used in policy-making as much as in everyday life (including the uncertainties, goals, and issues of self-presentation which present themselves to private individuals). They are a way of making sense of an uncertain, complex and contested world. In a more strategic sense, narratives may also be a means of persuading others, and this is particularly important in the policy process. In no way is the labelling of an account as a 'narrative' meant to be derogatory or to imply falsehood or fantasy. On the other hand, however, we cannot assume that we know the actors' intention from our interpretation of what they say. Also, narratives serve to stabilize their expectations and provide secure moorings in a shifting and sometimes threatening world, but they also perform representative and political purposes in the exercise of power by persuasion. Narrative analysis is therefore well suited for the treatment of policy (Roe 1994; 1995; Hajer 1995; Apthorpe & Gaspar 1996; Apthorpe 1997; Forsyth 2003; Springate-Baginski & Blaikie 2007).

Finally, the uses (and abuses) of policy narratives for purposes of understanding and promoting policy reform need to be recognised. A focus on narrative in the policy process is illuminating but insufficient by itself without addressing the reflexive relationship between the power of the author of the narrative and the narrative itself. This relationship may have structural aspects, such as funds for research and to command information, the means of broadcasting the narrative, networks of allies, and the means of coercion to produce and publicise the narrative as well as the persuasiveness – and therefore the power – of the narrative itself. This is particularly relevant to the Sri Lankan situation, where there are marked inequalities in the means to assemble and disseminate narratives to powerful audiences that may be persuaded to contribute to political leverage, funds and arms to the narrator's cause. Thus, it is not only the power of the narrative itself (the ability to persuade target audiences) but also the power of the author deriving from access to funding, military power and surveillance of civilian populations. In terms of the conflict in Sri Lanka, access to the media and to international funding agencies or international funding institutions (IFIs), as well as the power wielded by donors that comes with money, gives a narrative a captive audience. What happens subsequently, as the GoSL, NGOs and other actors in the field interpret and implement policies, may be a completely different matter. There are examples in this Issue (particularly by Hyndman 2009, and Lang & Knudsen 2009) of the GoSL's ability to prevail over a range of international and national institutions in terms of the narratives which are installed as official policy concerning Internally Displaced Persons, protection and humanitarian aid, as well as their implementation at the local level. Therefore a focus on policy narratives alone will tend to privilege documents and printed matter and also the formal and well financed over the spoken narrative of those who have poor access to a printing press, radio or politically powerful actors at different levels – an old story, but here in a contemporary setting. Other accounts of preferences – or in more urgent terms, acute needs – of the very people whom policies are supposed to be helping are frequently not heard or are buried by official narratives. Tete (2009) talks of subaltern narratives from certain sectors of IDPs (focusing on the right to 'return home' and the violation of IDPs' sense of place and home). This narrative contradicts notions of the 'territorialised state' which is strongly influenced by the strategic and cultural claims of the majority ethnic group represented by the GoSL. In this Issue, other narratives, both official and subaltern revolve around issues of protection, security, internal displacement, livelihoods (John Nigel 2009), housing (Brun & Lund 2009, this Issue) and land zoning (Hyndman 2009).

As in most policy-making, multiple actors, often with divergent versions of the 'truth' and competing objectives, are involved in negotiating formal policy. In many cases, outcomes of policy and what happens on the ground may bear little resemblance to the intentions of those who shape and draft policy documents. Some authors have taken a highly sceptical view of the conventional rationalist view of the policy process. Apthorpe (1997), for instance, says that 'the plainer and clearer a policy is painted, the more it is driven by evasion and disguise'. While our approach to policy acknowledges evasion and disguise, it is also tempered by an acceptance of the existence of instrumental intentions of the authors of policy in which there is often good faith, professionalism and a rational application of policy, as they see it, towards achieving stated outcomes. In the case of Sri Lanka after the tsunami, there were a large number of actors who were obliged to negotiate with the GoSL, the latter taking a centralised role, especially regarding the strategically sensitive issue of the distribution of humanitarian relief between the combatants. As Lang & Knudsen (2009) discuss, the policy concerning tsunami-related IDPs was highly contested and the humanitarian agencies faced very difficult political situations in their negotiations with the GoSL. Opposing narratives of best practice, practical necessities of protection, and the need for a strong centralised and coordinated relief effort (overseen by the GoSL) were all given some space, although particular forces within the GoSL were able to insist on recognition of some of the main points of their political agenda. Other voices may not be heard and be unable to join the negotiating table, or their knowledge may be deemed by more powerful actors to be worthless or illegitimate. The importance of listening to voices that do *not* contribute to formal policy-making may sound like a contradiction in terms (e.g. fishing communities, ethnic minorities who do not speak the language of the government, orphans, women in both rural areas as well as in bureaucracies, and other subalterns whose narratives are already mentioned). Another discussion of a narrative in a different policy area concerning IDPs is given by Brun & Lund (2009), who examine a particular housing narrative of 'One Size Fits All', which has generally excluded the voices of IDPs who speak of their own housing needs. The durability of narratives such as this one, even in the face of contrary evidence and repeated policy failure, is a common characteristic of certain narratives which perform well in terms of making sense of complexity, reducing uncertainty, appealing to 'common sense', and in policy terms are implementable and 'do-able' (see Scott 1998 for an extended discussion). The 'One Size Fits All' narrative for housing policy has proved its usefulness to successive administrations and resettlement projects for these reasons, but it has also proved to have very serious flaws which surface again and again, which the authors expose in their article.

Thus, there is a need to take into account the decisions and actions being made by others on the ground *outside* the formal policy process. Although formal documents, written MOUs (memoranda of understanding) and the narratives from which they may be constituted are essential texts to be read as a part of an explanation of a policy outcome, processes outside the formal policy process may well be as important. These actions also help to shape a de facto policy on the ground, but may only affect the formal policy process as a reaction or response to a previous cycle of policy-making. Simply, all manner of outcomes may unfold on the ground, quite unrelated to formal policy processes. The articles in this Issue illustrate some of the main aspects of the policy process itself – the evasions, good faith, ambiguities, and contested strategies of the main actors. Lang & Knudsen (2009) examine the actual operational processes at the interface between protection policy and practice in the context of Trincomalee District. They examine the challenges of humanitarian agencies operating in a difficult and insecure environment in which international normative frameworks for approaching internal displacement confront the political and military goals of the government of Sri Lanka. They also discuss some of the obstacles of operating in a context of internal displacement where the primary measures and responsibilities for maintaining civilian security rest with the political and military actors that are party to the war, and they sketch some of the creative strategies crafted by agencies operating in the field to counter these challenges. Brun & Lund (2009) show how the Divisional Secretariat and other civil servants at Divisional level emerged as important actors who played key roles in the way housing policy was played out on the ground. For a more generic literature on policy narratives, see Apthorpe & Gaspar (1996), Scott (1998), Keeley & Scoones (1999), Sutton (1999), and Shankland (2000).

The examples in this Issue illustrate the importance of understanding the context in which actors and their narratives work and have their day-to-day lives in terms of the structures, institutions and the relationships in which the actors are involved. Policy is not made only by political leaders in conjunction with the senior bureaucrats of the ministry concerned, but is profoundly affected at all stages by a whole cast of other actors including other ministries, IFIs, 'street-level bureaucracies' (i.e. the field staff at the lower levels of the policy process), the judiciary, and also by civil society (e.g. social movements, NGOs, prominent scientists and intellectuals, local politicians, state politicians and entrepreneurs). Silva's (2009) comparison of the malaria epidemic in 1934–1935 and the tsunami of 2004 exposes the importance of widening the analysis to take account of a much broader cast of actors operating at different levels. A key idea here is the 'politicised organigram' (organisation chart), which breathes real life in 'what really happens' in formal administrative structures. This is illustrated by Lang & Knudsen (2009), where they discuss the relationships between mandated and non-mandated institutions and the GoSL and also the relationships between international and local agencies on the ground. This article reviews humanitarian protection in practice, in terms of the place, positions, approaches and capability of diverse humanitarian agencies to negotiate politically sensitive relationships with the GoSL. Hyndman (2009) also discusses the ways in which policy plays out in the political geography of displacement in Sri Lanka. She shows how in Aceh, Indonesia the particular responses to the tsunami accelerated and facilitated the peace process, while in Sri Lanka, a coalition of nationalist politicians, the Sri Lankan military and a variety of local interests upended the Post-Terminal Operational Management Structure (P-TOMS) MoU and turned its back on the opportunities of 'cash for peace' and the 'peace dividend' which the international community had offered.

## Policy narratives about tsunami and civil war, and the actors who tell them

### *Vulnerability*

The concept of vulnerability has been in academic circulation for more than twenty years (since the discussion in Hewitt 1983). It has already been used in discussions about the effects of the civil war and the tsunami (see particularly John Nigel 2009 and Khasalamwa 2009). A major criticism which has strong practical application has been that the notion of vulnerability has also brought with it assumptions concerning the 'victims' of disasters, and their powerlessness. This has encouraged outside agencies involved in humanitarian assistance to underestimate the agency of the people they set out to help, and to fail to understand (or even attempt to find out) what the affected people can do themselves to mobilise their own resources and networks, with the result that huge funds are mobilised and wasted on the purchase of inappropriate, unacceptable, and unneeded supplies of food and temporary shelter (Blaikie 1994; Wisner et al. 2004). Lack of funds has seldom been the problem in immediate post-disaster situations and humanitarian relief in Sri Lanka and elsewhere. A total of USD 4.5 billion was pledged by the international donor community, and a proposal to organise the distribution of this sum was made (the Post-Tsunami Operational Management Structure (P-TOMS)). While international funds, food and the individual donations of a well-wishing and generous world public were allowed to flood the scene, the proposal caused protracted political acrimony and was never adopted, allowing disorganisation and ethnic discrimination to confuse the humanitarian and recovery effort (Hyndman 2009). Three major criticisms have been made of the recovery effort: first was the failure to address the complexity of vulnerability in a dynamic and fast-moving situation; second, the failure of not consulting the affected people themselves, listening and interpreting their own narratives, and involving them in their own recovery; and third, the susceptibility of programmes to centralised and politicised manipulation, as already mentioned. The failure to allow and create processes for the voices of the intended beneficiaries to be heard and to invite them to shape policy-making and implementation is a recurring theme in this Issue (Brun & Lund 2009; Khasalamwa 2009; Tete 2009). However, such a criticism, made after the event and often by those far from the theatre of day-to-day urgencies, should carry with it some humility, too. It is useful to bear in mind that the Sri Lankan effort at providing immediate humanitarian aid to the tsunami-affected population and subsequent reconstruction, for all its flaws discussed in this Issue, compares favourably in a number of crucial aspects to the humanitarian and recovery efforts following Hurricane Katrina and the disaster in New Orleans in 2005. From an academic viewpoint, a vulnerability analysis has tended to be a static categorisation of people by gender, ethnicity, wealth, and location (where people find themselves in hazardous places) rather than a process-oriented model which instead asks questions about how people adapt, cope, invent new strategies, forge new networks – and exploit or abandon others in a fast-moving military and political situation. Nonetheless, vulnerability has remained one of the key narratives of post-disaster recovery and disaster preparedness (see Khasalamwa 2009, who links the need to reduce vulnerability to the Principles and Good Practice of Humanitarian Donorship, and also John Nigel 2009). However, vulnerability (as also the narrative of livelihoods, discussed in the following section) has principally been developed in universities, INGOs and bilateral donors (e.g. the UK Department for International Development). It is complex and links a wide range of different policy sectors, and requires a deep political, social and economic understanding and an appreciation of the process of the reproduction of vulnerability which often is rooted in a long history. Implicit, too, in the vulnerability narrative is the humanitarian principle of the moral necessity for a reduction of damage, loss and suffering. Since it is a narrative primarily promoted by INGOs, NGOs and 'outsiders' to the GoSL, it has to engage with national and local politics freighted with inter-ethnic hostility, and also military and strategic considerations.

### *Livelihoods*

In the last ten years there has been an evolving set of models, collectively called the 'livelihoods approach', initially applied to rural development and later to post-disaster crisis and reconstruction. They have offered considerable advances to the vulnerability approach. Livelihood approaches complement and often overlap with vulnerability approaches but the former acknowledge and model peoples' own strategies in a dynamic framework more explicitly – it is a matter of what people can actually *do* in the face of threat within the different resources available to them at the time. Korf (2004) has applied a version of the model to the war situation in Sri

Lanka to highlight how farmers cope and adjust their economic and political assets to meet the threats made by a rapidly moving war situation. He also makes the point that war presents threats not only to livelihoods but also creates opportunities, which are taken up by local leaders who have advantageous ethnic connections with the army to bring about new and permanent economic gains. John Nigel (2009) has also reviewed the sustainable livelihood approach to the Sri Lankan situation and links the threats to livelihoods to the daily life of people caught in or near the fighting.

An understanding of a dynamic livelihood analysis (linked closely to the issue of vulnerability) is both essential to effective policy-making for humanitarian and development practitioners involved in both the tsunami and the civil war for post-tsunami and post-conflict situations. However, *understanding* the problems, applying experience from previous similar events and developing new theoretical models, while essential steps, are far from sufficient for *solving*, or even ameliorating these problems. The focus then shifts from understanding international narratives such as vulnerability and livelihood analysis to how they succeed or fail to reform policy in the widest sense – in this case the policies of the GoSL and the other agencies involved in humanitarian assistance and long-term rebuilding and recovery. This new focus shifts from the academy and claims to rational model building and policy formation to the sphere of the policy process. Hyndman (2009), offering an extended comparison between post-tsunami policy in the shadow of war between Sri Lanka and Aceh, convincingly shows that the former was (and continues to be, since the abrogation of the ceasefire in 2008) once more overshadowed by strategic and political considerations of the war – whatever narrative of reducing vulnerability, improving livelihoods or good humanitarian practice is on offer. Policy, indeed, is politics under another name. It is abundantly clear that policy which focuses on long-term recovery and development in the wake of the tsunami will continue to remain in the shadow of civil war.

As in the case of vulnerability as a policy narrative, the 'livelihoods approach' is also a product of international institutions, bilateral aid agencies and universities. In the Sri Lankan case, the actors who promote this narrative for long-term recovery from both the tsunami as well as the conflict tend not to be national but international (for example, CARE International, Oxfam and UNDP (United Nations Development Programme), as John Nigel (2009) discusses. Therefore the most significant interface (and often barrier) to the implementation of this approach as it is envisaged is that of national governments, which often invoke the principle of sovereignty in order to keep the livelihood approach and the international agencies which promote it at arms' length.

## *Reconstruction and recovery narratives*

There are a number of linked policy narratives which address issues of recovery and reconstruction after the disaster event of the tsunami and, more optimistically and uncertainly, after the cessation of the civil conflict. These are discussed at length in this Issue.

The IDPs 'right to return' is a central policy narrative in Sri Lanka in the context of both the tsunami and the civil war. It is essentially a restatement of human rights in a framework of what are called 'durable solutions' (Tete 2009). These involve IDPs right to return to their place of origin, their local integration and resettlement, and they uphold IDPs' rights and guarantees. It is the responsibility of the national government to provide the necessary resources and facilities that enable these durable solutions to be achieved. Tete examines the practical and political processes which unfold as the narrative leaves the paper on which it is written and proceeds across the interface between international actors and the national government and to INGOs and NGOs which may be involved in attempting to implement policies which uphold the framework. The necessary informed choice made by the IDPs themselves is integral to this policy in principle and practice, and becomes crucial, yet according to Tete the IDPs are seldom given any space to express their choice. The desire for the national government to exercise sovereignty and control (amounting to surveillance of civilian populations, as a number of authors in this Issue claim) and to achieve political advantage through the selective adherence of this framework clearly works against the expression of a 'right to return' as expressed by IDPs themselves.

There are other important policy narratives discussed in this Issue. The 'Build Back Better' policy narrative (the phrase coined by President Clinton in 2004), again deriving from a coalition of international donors and organisations, is one central to a more general humanitarian discourse (Khasalamwa 2009). A central policy goal of 'Build Back Better' is to seize the opportunity which a disaster such as the tsunami affords and reduce vulnerability in the short term as well as in the longer term in view of possible future natural events (such as another tsunami) involving post-crisis recovery and Disaster Mitigation and Preparedness. The 'Cash for Peace' narrative developed by President Yudhoyono of Indonesia in the context of the civil conflict in Indonesia after the tsunami which devastated the coast of Aceh is a parallel case discussed by Hyndman (2009). In Sri Lanka, the interface between international and national policy formulation and implementation is enshrined in the Principles of Good Practice of Humanitarian Donorship. The National Housing Policy, discussed by Brun & Lund (2009), intersects with 'Building Back Better' but their focus is distinct in a number of important respects. The historical antecedents of a top-down, unvarying, 'one-size-fits-all' housing policy dates back to the 1930s, the authors claim. Maps and plans of settlement schemes attest to a blueprint where a variety of cultural and practical priorities of the people who were to live in them were ignored – and have been for the previous seventy years. Here, as in many other policy-making areas throughout the world, a blueprint satisfies many bureaucratic goals; it replicates a 'tried and tested' formula (not by residents, but by government officers), it reduces the complexity of different cultural and physical planning variables and does not run the risk of ceding control to local people. In addition, Brun & Lund (2009) argue, there are familiar ethnic and political biases in the Sri Lankan case and these are similar to those discussed in other areas of the policy process discussed in this Issue.

## Recurring themes

It may be useful in this Introduction to signal a number of themes which reoccur in this Issue. First, any framework which gives the opportunity to understand better the policy process in any context (not only in Sri Lanka) may claim that it is likely to encourage the prospect of reform – however remote this may be in the light of other political forces. This Special Issue does not claim to offer a new approach but it can claim to bring together a number of hitherto separate areas of policy research. It gives attention to the policy narrative itself – its frames of reference, assumptions, problem focus, what is in the frame and left out, its logical structure, its language and metaphors, and its ethical and ideological underpinnings. Then, linked to the narrative, the focus is turned to the actors who develop, negotiate, promote, struggle with, or subvert these policy narratives. The conceptualisation of power is always a sensitive task and in this Introduction and at different points in the Issue a partial and context-specific set of indicators are suggested. The national government and those small groups within it who can apply leverage at particular moments are key powerful actors against which other international ones (for example, charismatic world political leaders, bilateral and multinational agencies, and INGOs) have an episodic and fitful impact. The extremely delicate political negotiations with the GoSL with which humanitarian agencies have to grapple are discussed in detail by Lang & Knudsen (2009). This is not to underestimate the power to influence policy and outcomes on the part of the international actors, but rather to understand their leverage relative to senior politicians at the national level and also local level functionaries who may wield the power to shape outcomes decisively. An important focus upon the cast of actors and their interactions is maintained in this Issue upon those who are affected by policy decisions but whose voices remain unheard, or deliberately silenced. The narratives in this Issue on housing policy ('One Size Fits All'), the 'right to return' of IDPs, and humanitarian protection of those threatened by civil conflict, are three examples where the voices of affected people in the greatest need are seldom heard. As a follow-on from the identification of the actors involved, it is helpful to understand their day-to-day realities and their interactions regarding others as they shape policy on the ground. Related to this is a necessary realisation that the authors who write in this Issue are also actors, albeit usually minor ones, often with no more than a 'walk-on' part, without a script and lines to speak in the policy process. Usually, those authors who have other more 'hands-on' and direct roles in the policy process will have been involved in shaping policy more definitively in roles other then solely academic. Tracing the impact of academic articles to changes in policy usually involves a difficult and often optimistic judgement.

Lastly, the policy process is, as the phrase implies, one of continuing unfolding. There is an ongoing reflexive set of relationships between actors and between their narratives. Some treatments of the policy process assume that *all* is interim and policy is always an unfinished story. Certainly, the policy process constantly changes and all the articles in this Issue build a time element into their accounts of the policy process. As this Issue goes to press, the policy process will have moved on as will have the military situation. Nonetheless, some policy narratives (and most in this Issue) have durability. For example, 'One Size Fits All' has had a successful career of approximately seventy years, mainly because it has discursive power (and like many other enduring narratives it appeals to chosen audiences, has sweetness in the telling and resonates with dominant political and ideological aspirations of those with most power at the time). Others have a compelling moral certainty (for example, those which appeal to human rights), and others again have powerful patrons and authors. Thus, many of the policy narratives discussed in this Issue will endure, but whether this critical discussion on their deployment will lead, however indirectly, to reform at some future date remains to be seen.

## References

Apthorpe, R. 1997. Writing development policy and policy analysis plain or clear; on language, genre and power. Shore, C. & Wright, S. (eds.) *Anthropology of Policy: Critical Perspectives on Governance and Power*, 34–45. Routledge, London and New York.

Apthorpe, R. & Gaspar, D. 1996. *Arguing Development Policy: Frames and Discourses.* Frank Cass, London.

Blaikie, P.M., Cannon, T., Davis, I. & Wisner, B. 1994. *At Risk: People's Vulnerability and Disasters.* Routledge, London and New York.

Brun, C. & Lund, R. 2009. 'Unpacking' the narrative of a national housing policy in Sri Lanka. *Norsk Geografisk Tidsskrift–Norwegian Journal of Geography* 63:1, 10–12.

Forsyth, T. 2003. *Critical Political Ecology: The Politics of Environmental Science.* Routledge, London and New York.

GoSL. 2005. *Government of Sri Lanka Post-Tsunami Reconstruction and Recovery Strategy.* Government of Sri Lanka, Colombo.

Hajer, M.A. 1995. *The Politics of Environmental Discourse.* Clarendon Press, Oxford.

Hewitt, K. (ed.) 1983. *Interpretations of Human Calamity from the Viewpoint of Human Ecology.* Allen and Unwin, London.

Hyndman, J. 2009. Siting conflict and peace in post-tsunami Sri Lanka and Aceh, Indonesia. *Norsk Geografisk Tidsskrift–Norwegian Journal of Geography* 63:1, 89–96.

Keeley, J. & Scoones, I. 1999. *Understanding Environmental Policy Processes: A Review.* IDS Working Paper 89. Institute of Development Studies, University of Sussex, Brighton.

Kelman, I. & Gaillard, J.-C. 2007. Disaster diplomacy in Aceh. *Humanitarian Practitioners Exchange: Practice and Policy Notes* 37, 37–39.

Khasalamwa, S. 2009. Is 'build back better' a response to vulnerability? Analysis of the post-tsunami humanitarian interventions in Sri Lanka. *Norsk Geografisk Tidsskrift–Norwegian Journal of Geography* 63:1, 73–88.

Korf, B. 2004. War, livelihoods and vulnerability in Sri Lanka. *Development and Change* 35:2, 275–295.

Lang, H. & Knudsen, A. 2009. 'Your subject of protection is a dangerous one': Protracted internal conflict and the challenges for humanitarian agencies. *Norsk Geografisk Tidsskrift–Norwegian Journal of Geography* 63:1, 35–45.

Nigel, J. 2009. Livelihoods in a conflict setting. *Norsk Geografisk Tidsskrift–Norwegian Journal of Geography* 63:1, 23–34.

Roe, E.M. 1994. *Narrative Policy Analysis: Theory and Practice.* Duke University Press, Durham and London.

Roe, E.M. 1995. Introduction: Special section, development narratives. *World Development* 23:6, 1005–1006.

Schrijvers, J. 1999. Fighters, victims and survivors: Constructions of ethnicity, gender and refugeeness among Tamils in Sri Lanka. *Journal of Refugee Studies* 12:3, 309.

Scott, J.C. 1998. *Seeing Like a State: How Certain Schemes to Improve the Human Condition Have Failed.* Yale University Press, New Haven, CT.

Shankland, A. 2000. *Analyzing Policy for Sustainable Livelihoods.* IDS Research Report 49. Institute of Development Studies, University of Sussex, Brighton.

Silva, K.S. 2009. 'Tsunami third wave' and the politics of disaster management in Sri Lanka. *Norsk Geografisk Tidsskrift–Norwegian Journal of Geography* 63:1, 61–72.

Springate-Baginski, O. & Blaikie, P.M. 2007. *Forests, People and Power: The Political Ecology of Reform in South Asia.* Earthscan, London.

Sutton, R. 1999. *The Policy Process: An Overview.* ODI Working Paper No. 118. Overseas Development Institute, London.

Tete, S.Y.A. 2009. Whose solution? Policy imperatives vis-à-vis internally displaced persons' perceptions of solutions to their situation in the Sri Lankan conflict. *Norsk Geografisk Tidsskrift–Norwegian Journal of Geography* 63:1, 46–60.

Wisner, B., Blaikie, P.M., Cannon, T. & Davis, I. 2004. *At Risk: Natural Hazards, People's Vulnerability and Disasters.* Routledge, London and New York.

# 'Unpacking' the narrative of a national housing policy in Sri Lanka

CATHRINE BRUN & RAGNHILD LUND

The article 'unpacks' the major policy narrative on housing in Sri Lanka, namely that there is *One National Housing Policy* (ONHP). Housing schemes have been part of poverty alleviation strategies designed to solve problems of internal displacement due to war, and more recently as part of the major recovery initiatives after the Indian Ocean tsunami event in 2004. The housing policy narrative is explored in a resettlement programme (1970s to date), in a case of war-induced internal displacement (1990s – early 2000), and in post-tsunami recovery (from 2005). The authors ask whether successful practices from the past have been incorporated into the post-tsunami recovery efforts. Although similar in structure and implementation processes – all highly centralised, technocratic, bureaucratic, and top-down – each scheme has its own policy narrative which is based on different local contexts and experiences. Housing plans are never neutral, but embedded in existing situations of tension and are highly politicised. New to the post-tsunami situation is that the housing policy has not sufficiently embedded the reconstruction practices in local realities and people's own preferences and contributions. Ignorance among international organisations about previous housing policies and power relations has led to a silent acceptance of the ONHP.

*Cathrine Brun, Department of Geography, Norwegian University of Science and Technology*
*Ragnhild Lund, Department of Geography, Norwegian University of Science and Technology*

## Introduction

When Sri Lanka had to 'rebuild' communities as a consequence of a catastrophic tsunami hitting the coast of Sri Lanka on 26 December 2004, housing became a major issue in terms of both relief (providing shelters and temporary houses) and reconstruction (constructing new and rebuilding damaged houses). Subsequent to the tsunami waves, a 'wave' of aid workers, international consultants and other international actors with international funding flooded the country. A relatively efficient relief stage by local people and others was rapidly replaced by politically complex intermediate and recovery stages.

This article[1] focuses on how policies and practices must be understood in the context of a narrative that has been formed by various actors over decades of housing and settlement practices in the country. By understanding how actors contribute to the policy narrative and, consequently, how the policy narrative frames the post-tsunami housing and settlement interventions, we attempt to understand, first, past experiences and practices of housing and settlements in Sri Lanka, and second, how past experiences are monopolised by one dominant national housing policy narrative which has restricted the opportunities to learn from past experiences.

We are mainly concerned with the policies and practices of housing, and thereby contribute to discussions on why the post-tsunami housing programmes have been a mixed blessing in Sri Lanka. We do not attempt to give a comprehensive analysis of these programmes. Also, we are looking mainly at past experiences in housing policies that may provide an understanding of government policies in Sri Lanka. This perspective could be part of the explanation for the magnitude of the tsunami disaster, as Kennedy et al. (2008) mention: poor planning and coastal zone regulations had contributed to rapid population increases in coastal areas which increased vulnerability to tsunamis. While this relates to poor planning and/or lack of intervention by the Government, in this paper we are more concerned with the history of planned housing interventions in Sri Lanka. Two key dimensions of the One National Housing Policy (ONHP) narrative are discussed: first, that housing, understood as building shelter and forming settlements, is a meaningful way of 'doing' development and, second, that there is a strong emphasis on the physical dimensions of housing related to the way houses should be built and settlements planned, in which people should participate. There is a need to 'unpack' the knowledge and practices embedded in the ONHP narrative, and explore whether we can uncover multiple meanings that can challenge this narrative. We ask: What guides the development and implementation of the national policies relating to settlement and housing? What, if any, lessons have been learnt from previous schemes? What ideas and policies guide them, and make them follow similar paths of development? What is the knowledge and what are the practices embedded in the narrative of the One National Housing Policy?

'Lessons learnt' from empirical cases – such as the Mahaweli project in the Central Province, the housing schemes for internally displaced people (IDPs) in Puttalam in the North Western Province, and the post-tsunami housing schemes in various sites in Sri Lanka – are juxtaposed. Relying on our own previous research in these areas, we aim to show how, during various types of development projects and crises, the Government of Sri Lanka has pursued different housing strategies with variable success. Even if the country has had a long tradition of settlement and housing schemes, is it the case that Sri Lanka follows one national blueprint policy and model, or was the tsunami situation unique, representing a break with former policies?

The sources of information used in this article are based on insights gained from our previous ethnographic studies in the Mahaweli area and Puttalam (Lund 1981; 1983; 1993a; 1993b; Brun 2008; Brun & Lund 2008). During the period 2005–2008 we conducted desk studies on needs assessments and INGO (International Government Organizations) livelihood practices, in-depth research on post-tsunami housing schemes in the east, north and south of the country, and interviews with people who had lost their house and were in receipt of housing from INGOs and other support from community-based organisations (CBOs). In addition to our own archives, secondary sources included policy papers and evaluation reports.

## The narrative on one national housing policy

The dominant and 'persistent' policy narrative on housing in Sri Lanka today is that there is *One National Housing Policy* (ONHP) which has existed for as long as settlement programmes have existed in Sri Lanka, starting with the early colonisation schemes after Sri Lanka gained Independence from British rule (in 1948). Since then, housing schemes have been part of poverty alleviation programmes and other strategies to solve problems of internal displacement due to war. Such schemes have been strategically important for various governments and political leaders, and policies, models and technologies relating to housing have been readily available in the country.

The tsunami disaster of 26 December 2004 is historically one of the most extensive disasters to have affected Sri Lanka, but it was not the first. Silva, for example (2009, this Issue) argues that other disasters have been far more devastating to the country, such as the malaria epidemic of 1934–1935, which lead to death tolls twice to that of the tsunami: officially, 40,000 were killed by the tsunami, while the malaria epidemic recorded 80,000 dead. Another example is that of the cyclone which affected Sri Lanka in 1978, and hit 20% of the country and affected 800,000 to 1 million people, causing deaths, injuries and severe floods (Lewis 1981; 1999). Still, there has been a tendency to look at the tsunami event as exceptional and the recovery process as detached from certain local, national and international experiences. However, although the tsunami was an exceptional disaster, the recovery stage has brought with it challenges that have been experienced in other housing and settlement schemes in Sri Lanka.

Policy can be defined as a set of official actions carried out by an institution, often governmental, in the name of the public good. Policy narratives usually include argumentation, and sometimes assumptions and production of evidence, upon which policy is based. Such narratives may be real or they may be myths. They may co-exist and some may emerge as dominant and thus gain power over others. When we compare the narrative on housing in Sri Lanka over time, many similar practices are revealed, even though the story and the language shift over time. Still, many would argue that after the tsunami a new language was created for housing, related to the uses, contexts, timing, and debates.

In the words of Cornwall et al. (2007, 2), a narrative is about 'the creation and evolution of new languages – languages of representation, languages of analysis and languages of policy discourse – and debate over these'. According to Blaikie et al. (2007, 93), 'policy narratives are stories told by different protagonists . . . as a way of making sense of an uncertain, complex and contested world or more strategically, as a means of persuading others'. While these definitions are concerned with the language and stories of policy narratives, in this paper we are equally concerned with how policy narratives tend to shape people's practices for solving a particular challenge in a particular way. We consider policies, their interventions and the narratives created in this process as an ongoing socially constructed and negotiated process (Long 2001). Policy narratives are made by past policies and interventions, the way actors organise and strategise around these policies, and hence the outcome of the policies. When we explore the importance of the ONHP narrative by juxtaposing the practices of two historical cases – Mahaweli and Puttalam – with the practices of post-tsunami housing reconstruction, for example, we learn how knowledge of past experiences is embedded in the dominant narrative. The ONHP narrative is manifested in the maps, plans and physical features of housing. People's own stories of building homes and continuing with their everyday lives are not significant parts of the narrative. In order to unpack the narrative, we follow Long's (2001, 72) understanding of planned intervention:

> [P]lanned intervention is a transformative process that is constantly re-shaped by its own internal organisational, cultural and political dynamic and by the specific conditions it encounters or itself creates, including the responses and strategies of local groups who may struggle to define and defend their own social spaces, cultural boundaries and positions within the wider power field.

Thus, the interface between the various actors in planned interventions is usually complex and it is difficult to give full account of the different ends and means attributed to different actors and their perceptions and expectations. In order to understand the working and importance of the housing policy narrative, we study the role of three sets of actors: government, local and national, the international donor community, and the people (and their organisations), for whom the housing was intended. While all have different perceptions and expectations about what housing can achieve, we also show through the three cases discussed here that the ONHP narrative is carried over from one housing programme to the next, and becomes embedded in the way various actors think about and practise housing in Sri Lanka.

## Housing as a development strategy

During the 1950s and 1960s, major resettlement schemes and land reform in Sri Lanka were part of national policies to improve the living conditions for the poor, landless and marginalised. Such schemes, in which the provision of housing constituted a major component, were developed nationwide, but particularly in the Southern, North Central, and Eastern Provinces. Later, in the late 1970s the country launched the Mahaweli project (which is still operational) as a major effort to resettle people in the North Central and Eastern Provinces (Fig. 1). This project has involved resettling thousands of poor people in need of farmland, infrastructure and housing. The construction of simple, one- or two-room houses on 0.5 acre (0.2 ha) plots, the provision of tools and building materials for self-construction and maintenance, and also the provision of user rights to additional farmland have all been essential components of this and other schemes, such as building houses for IDPs. In most schemes, the same blueprint models of nucleated settlement patterns have been implemented, leading to densely populated housing units, which are very different from traditional rural housing in small compounds, and urban housing in rows (cf. Figs. 2, 4 and 5).

However, such government-built schemes have faced serious legal problems due to lack of proper tenure arrangements for the occupants, as well as ethnic and political biases, and also the fact that the implementation and maintenance of the programmes have faced logistical problems (Lund 1993b; Mueller & Hettige 1995; Sørensen 1996; Scudder 2005). As will be shown in the next section, these challenges are today being replicated in post-tsunami housing efforts.

## *The Mahaweli project*

Until the initiation of the first colonisation schemes in the north-central region of Sri Lanka in the 1930s, this part of the country had long been isolated. In 1963, the Government, with the assistance of the Food and Agriculture Organization and United Nations Development Programme, launched the first mapping of the area in preparation for a huge development project in the Mahaweli river basin. It was found that the area had enough water, land and space to cater successfully for a resettlement programme of vast magnitude. Since then, the Mahaweli project, which covers about one-third of the country's area (Fig. 1), has played an important part in the country's development policies, as a strategy for solving some of Sri Lanka's demographic and economic problems (Lund 1993b). Here, we will particularly draw upon experiences relating to resettlement and housing in the first part of the construction which took place during the period 1978–1984, and which is termed the Accelerated Mahaweli Project. The project was to be completed within a time span of six years instead of thirty years as originally planned. Revisits to this area have taken place repeatedly

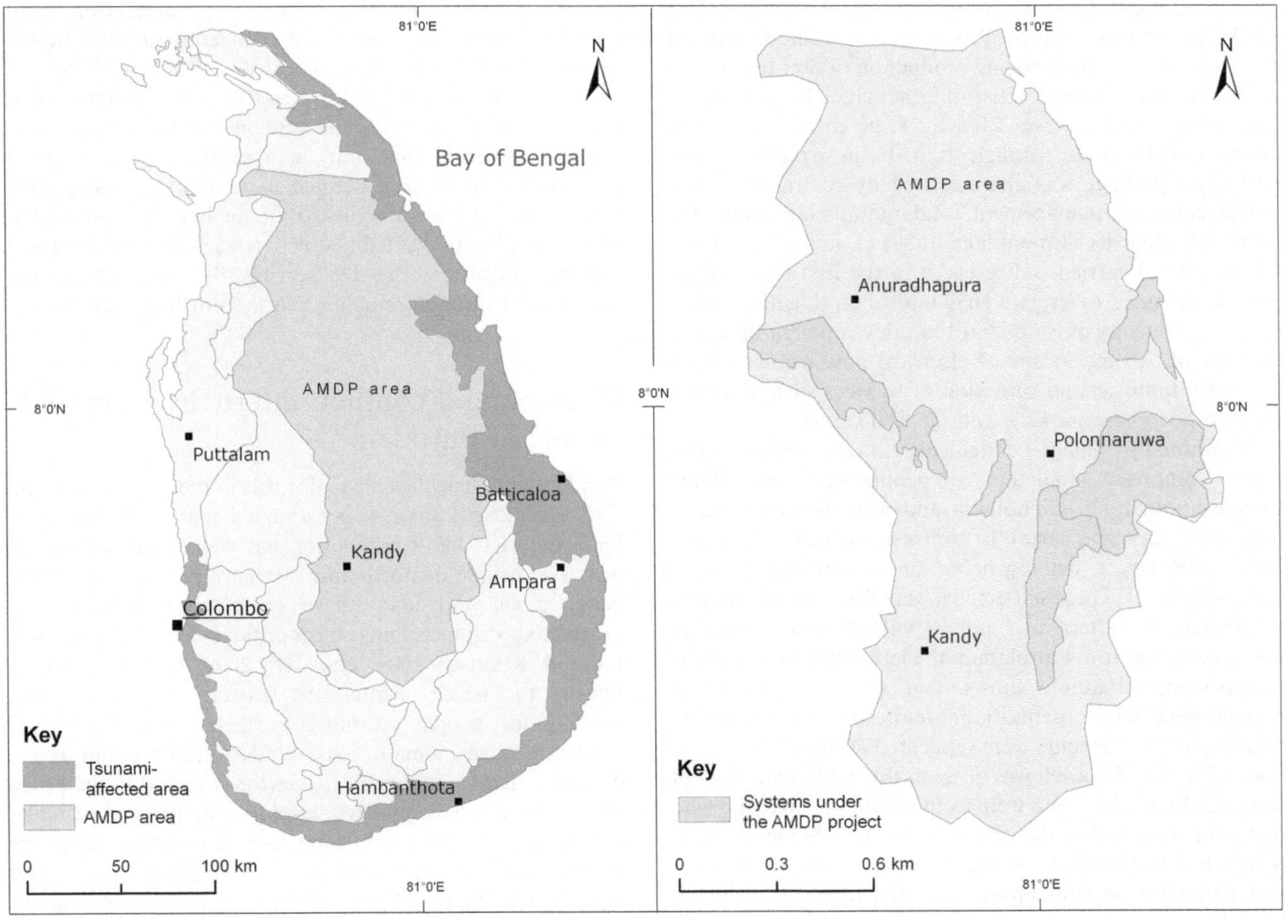


*Fig. 1.* Tsunami-affected areas and the Accelerated Mahaweli Development Programme area

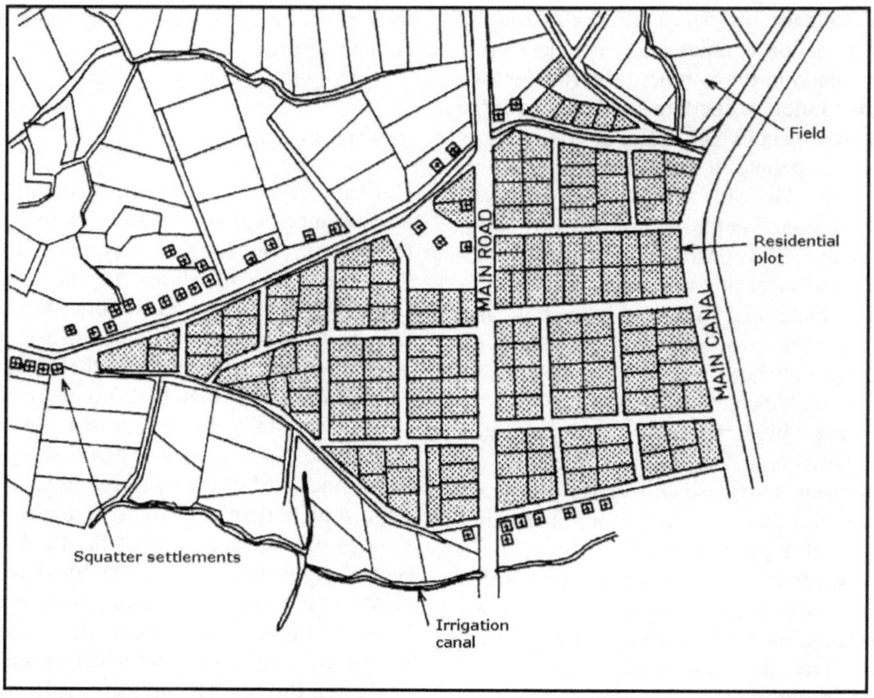


*Fig. 2.* Sketch plan of a settlement in Mahaweli illustrating a densely populated nucleated settlement pattern

since then, and the most recent visit was in 2004. The aims of the Mahaweli Development Project were to achieve national sufficiency in rice, to stimulate production of non-traditional crops, to create income generating activities in agriculture, and to construct a power capacity large enough to enable intensive cultivation through trans-basin irrigation. The settlement package was characterised by central planning, implementation, management, and monitoring under the Mahaweli Board (Karunatilake 1989). Lund (1981; 1983; 1993a; 1993b) carried out research in the first stage of the project, designed to irrigate 1659 km[1] of land, with a yearly irrigation capacity of *c.*7 milliard $m^3$, development of a total capacity of 500 megawatts of electrical power, resettlement of 75,000 families, and provision of services such as roads, schools, hospitals, markets, and medical centres.

Traditionally, village settlements in the north-central region comprised small units of people who were related to one another (20–25 houses) and were homogenous in nature (by class and caste). In contrast, the new settlements were much larger, heterogeneous units (100–125 or more houses) (Fig. 2). These settlements were formed according to a hierarchical pattern of hamlets, villages and townships. Later, evidence from Puttalam in the late 1990s as well as the post-tsunami situation shows that the same nucleated management and organisational pattern with respect to housing and settlements were replicated (Brun 2008).

Land in the Mahaweli was given to three different types of settlers: a) *resettlers*, who were from within the area and who had been resettled under the new scheme. Some of these people had lost land due to the project's construction works and had received compensation in the form of cash and land; b) *new settlers*, who were landless people from various parts of the country (many from the southern, heavily populated areas; and c) *Evacuees*, who came from land inundated as a consequence of dam construction. In the project, each family was allocated 2.5 acres (1 ha) of irrigated fields and 0.5 of an acre (0.2 ha) of home-garden land. Tools and raw materials were provided to set up a one-room 'temporary' mud hut on a self-help basis. Even if individual settlers were defined as landholders, they were actually only given the rights to *use* the land, not ownership of it.[2] This has prevented people from selling their land and moving elsewhere, thereby serving the interests of the government through remaining in the north-eastern region.

## Housing and conflict: internally displaced people in Puttalam

Displacement has played a major role in the discourse on the civil war in Sri Lanka. It is estimated that more than one-tenth of Sri Lanka's population has been displaced by the war. The people of north and east have paid a particularly heavy price, and many have experienced multiple and protracted displacements during their lifetime (see also Lang & Knudsen 2009, and Tete 2009, both this Issue). During the peace negotiations between 2002 and 2004, housing for people returning to their original areas of residence became a major issue. Under the concept of 'peace dividend', it was important for parties involved in the peace process to create a positive development process to make people feel they benefited from peace. Also, before the peace process and after the resumption of war, there was a strong discourse on the possibility of return for the people living in protracted displacement situations. These are situations that have required more permanency in how to deal with shelter while people are displaced. One situation of protracted displacement is the Muslims from the north of the country living as internally displaced people in Puttalam District. They were evicted from the LTTE (Liberation Tigers of Tamil Eelam) controlled areas in the Northern Province in 1990 and most people fled to Puttalam, where the majority stayed in displaced people's camps (Fig. 3).

A change of government policies in 1995 included a shift in policies towards IDPs nationwide, and enabled an opportunity for some people to move out of the camps and into more permanent settlements. While some assistance was provided by international NGOs to such camps and settlements, government policies have played a crucial role in the way camps are run, the formation of settlements, how houses were built, and consequently how displaced people strategised in relation to policies and implementation of such policies. Crucial to housing for IDPs displaced by conflict was the 'Unified Assistance Scheme' (UAS), a resettlement and relocation package initially implemented to assist the return of displaced people to their original places of residence. Under the extended UAS, displaced people could relocate and receive a form of multi-stage assistance. In the first phase, assistance to build a temporary hut was given, together with continuing food assistance. In the second phase, assistance was given in instalments to build a permanent house. The final stage of the package was a Productive Enterprise Grant, to enable a family to become independent of continuing food assistance. The formative government policies developed in 1995 carried with them the housing narrative developed from the Mahaweli project and other previous housing schemes, both regarding the importance of housing for people's development and the physical layout of the settlements (Figs. 4 and 5). What was new in the case of Puttalam was the active role people played in organising themselves to buy land and build settlements. Whereas under other housing programmes the government provided land, in this case people had to buy land themselves, as the government would not support the permanent shift of Muslims from the north. Most people bought very small plots, often only 10 perches ($c.$250 m$^2$). The size of these plots prevented them from carrying out any farming activities, and they did not have any land to pass on to their children, which reduced the sustainability of the programme.

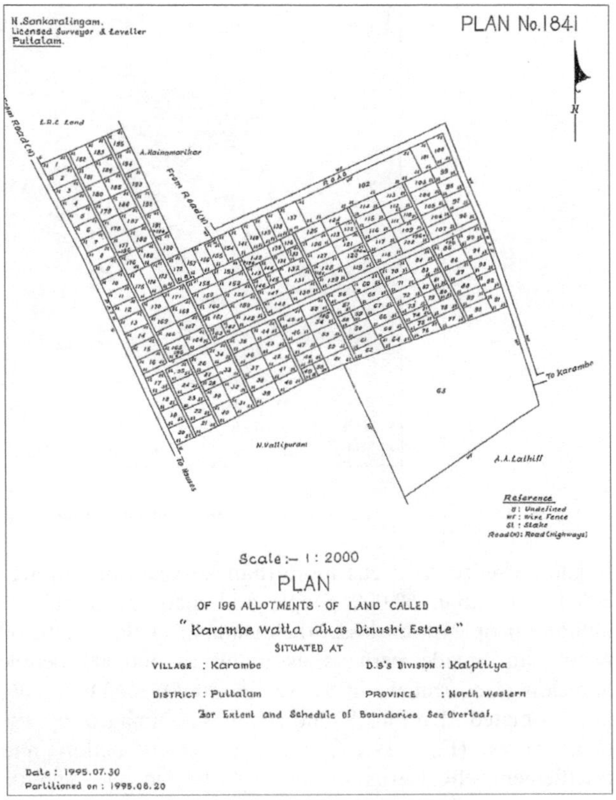


*Fig. 4.* Surveyor's plan of a new settlement in Puttalam, illustrating a densely populated nucleated settlement pattern (source: N. Sankaralinghma)

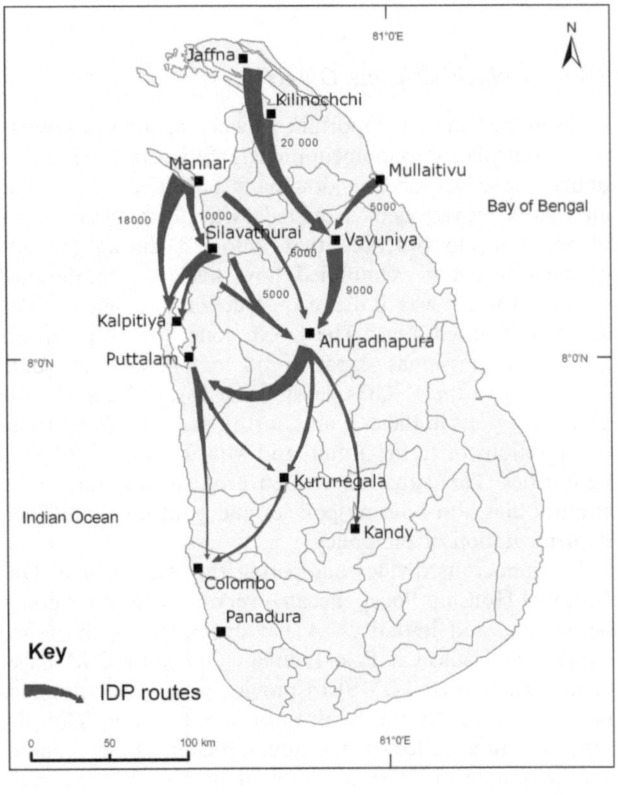


*Fig. 3.* IDP routes from the north in 1990 (based on Hasbullah 2001)

In Puttalam and other situations of conflict-induced displacement, relocation schemes have provided more permanent housing for the displaced and, as such, have made the lives of IDPs more bearable. However, at the same time, the schemes have resulted in people continuing to hope for return to the north and this has enabled the government to continue its strategy of control over the displaced people.

## Housing after the tsunami and the role of the state

When the tsunami hit the coasts of Sri Lanka, the country was almost on the brink of war after three years of a

*Fig. 5.* Poster of a post-tsunami housing scheme, illustrating high standard and density of housing

fragile ceasefire. Whereas more than 20 years of war had killed more than 60,000 people and made more than 1 million people homeless, the tsunami killed *c.*40,000 people and made perhaps as many as 500,000 people homeless (Government of Sri Lanka 2005). RADA (2006) has estimated that 50,783 houses were damaged by the tidal waves (Fig. 1). A massive reconstruction and resettlement effort was begun and the slogan 'building back better' (Clinton 2006; Steele 2006) became prominent in the discourse on post-tsunami reconstruction.

The government policy developed two main approaches to housing: owner-driven and donor-driven. In those cases where people stayed put and rebuilt their own partly or totally damaged house, they were given cash in instalments to carry out the work. Where people had to move to settlements away from the coast due to the buffer zone (which will be discussed in the following section), the donor-driven approach meant that the government selected land where the international organisations built houses. It is the latter process that we are discussing in detail here. However, in the beginning, the government and the collaborating INGOs formulated MoUs (Memoranda of Understanding) which included the division of labour between them, and at the same time framed institutional relations and responsibilities among them. The distribution and allocation of land was a central responsibility of the government, while the implementation of housing became the responsibility of the INGOs. The government also provided fixed standards of design for the houses, not allowing much freedom for alternative design. This confirmed a top-down process and ultimate control by the government.

This situation was not new to the top-down tradition of planning settlement and housing schemes in Sri Lanka. What *was* new was the heavy presence of international funding and international agencies, which created awkward relationships between the government and/or state and the international community in the execution of the policy processes following the tsunami. From the outset of the tsunami recovery work, shelter and housing were perceived as symbols of the devastation caused by the disaster. Providing housing was seen as a way to develop, to 'rebuild communities' and 'build back better'. It soon became clear that implementing agencies resorted to old paradigms on housing by mainly referring to physical housing reconstruction rather than thinking about facilitating change to create new homes and pro-poor development (Skotte 2004).[3] Claims were made by Government of Sri Lanka, and duly acknowledged by the international housing agencies, that there was a common, national housing reconstruction policy practised throughout the whole country after the tsunami. The national post-tsunami housing policy was clearly related to *housing as development*.

## *Biases embedded in the ONHP*

Housing became an important tool for creating a government-controlled development process (Silva 2009). However, under the cover of this dominant narrative, biases and alternative narratives have existed. After the tsunami, many of the housing practices that unfolded during the still ongoing recovery exemplified how weak or uninterested the Government was in the reconstruction of conflict-ridden and marginalised areas. The GoSL concentrated its efforts and its professional capacity on rebuilding the south together with the INGOs, leaving the harder hit and more isolated regions in the east and north to a plethora of donor organisations or to the Tamil- and Muslim-dominated local authorities. This situation created professional, ethnic and cultural biases in housing policies and guidelines, and their implementation of the policy.

The numerous divides hampering the efforts for a 'One National Housing Policy' became very clear in other policy discussions and initiatives. A key example was the failed efforts to establish a Post-Tsunami Operational Management Structure (P-TOMS) to provide island-wide humanitarian support to the victims of the tsunami. Housing reconstruction reflected the predominant ethnic conflict and geographical divisions. More than 3500 houses, more than the number damaged by the tsunami, had already been built in Hambantota District in the Southern Province, while only 25% of the damaged houses had been reconstructed in Ampara and less than 10% in Batticaloa, both Districts in the Eastern Province (RADA 2006; Perera 2007) (Table 1). It can be argued that this led to the total fragmentation of the housing reconstruction process. Regarding the distribution of houses, there was, in fact, no national housing policy at all; there may perhaps have been a national housing reconstruction *operations* policy, but no common housing policy as such because of the regional disparities that had developed.

The ethnic bias and identity politics embedded in housing policies have a long history in the country. For several decades, the Mahaweli project has been marketed as a way to bring Singhalese civilisation back to the north-central region of Sri Lanka. Various presidents over the years have tried to create populist ideology in an effort to legitimise Singhalese nationalist policies. To encourage people to move to the Mahaweli settlements, the authorities made use of various symbols and rituals, and legends and stories about the glory of the past, such as the 'Dawn of the Mahaweli Era', with the return of the people to their ancient 'homelands'. When dams, canals and new settlements were constructed under this project, temples were built and old rituals (e.g. inauguration ceremonies with oil lamps and prayers, certain thanksgiving rituals) were reintroduced (Lund 1993a). No similar processes of identity formation was created among the tsunami victims, who, like the previous Mahaweli settlers came from different villages and, in many cases, where resettled without knowing their neighbours.

Furthermore resettlement in the north-central region of Sri Lanka has contributed substantially to the changes in the population map of Sri Lanka. Politically and symbolically the project has implied resettlement of Singhalese people in previous Tamil areas, or close to them. For example, the large majority of the settlers in the Mahaweli were Singhalese Buddhists (David 1986; Scudder 2005). According to Scudder (2005), of the 75,000 households settled under the Accelerated Mahaweli Development Project (AMDP), only 1.9% were Hindu and 2.9% were Muslim. Some would argue that the project was a direct cause of the intensified ethnic tensions between Singhalese and Tamils in 1983. A consequence of the settlement schemes and movement of people was major shifts in demographic and ethnic composition, creating Singhalese majority in many constituencies which earlier had been dominated by Muslims or Tamils (Bastian 1996).

Similar trends of ethnicisation of the Sri Lankan society and polity are also clearly reflected in how, due to the war, the movement of people has both been caused by and has led to identity politics. For example, as mentioned above, in 1990, all Muslims from northern Sri Lanka were expelled by the Liberation Tigers of Tamil Eelam in the largest single case of ethnic cleansing in the history of the conflict. The displacees' decision to go to Puttalam was partly because it already had a majority Muslim population, and subsequently the movement of people has led to further ethnic segregation at both national and local levels. Still, in the case of Puttalam, the interest of the government and the leaders of the northern Muslims is that they should return to maintain claims to the north which is otherwise solely settled by Tamils. In 2007, President Mahinda Rajapaksa even ordered the northern Muslims to return. By maintaining the temporariness of the displaced northern Muslims, the Government is still controlling the population and using civilians in the civil war.[4]

According to our observations (September 2008), a similar ethnicisation process is occurring in the eastern region today, with the building of hotels, a port and other construction work controlled by the dominant ethnic group taking place in the buffer zone, while people who used to live there are being displaced to new settlements in the far interior. This has created increasing political and ethnic tension among the coastal population in a struggle for access to land in the densely populated Muslim and Tamil dominated areas.

## The ONHP and the role of the 'international community'

In the aftermath of the tsunami crisis much has been written on how donors carried out their work (Fernando & Hilhorst 2006; Cosgrave 2007), what was achieved and not achieved (Bennett et al. 2006; Braunmuhl et al., 2006; Saravananthan 2007), whose purposes were being served by external donors, and how the handling of tsunami funds in fact contributed to accelerate the conflict in the country (Uyangoda 2005; Hyndman 2007). These debates have led to a massive

*Table 1.* Houses damaged by and houses reconstructed following the tsunami

| District | Total houses damaged | No. units (MoUs) signed | No. units construction commenced |
|---|---|---|---|
| **Ampara** | **12,198** | **6716** | **3116** |
| Batticaloa | 5201 | 4591 | 313 |
| Colombo | 5150 | 1640 | 312 |
| Galle | 5568 | 2871 | 1924 |
| Gampaha | 326 | 377 | 243 |
| **Hambantota** | **1057** | **4073** | **4702** |
| Jaffna | 4508 | 439 | 411 |
| Kaluthara | 4275 | 2634 | 687 |
| Matara | 3752 | 2744 | 1242 |
| **Mullaitivu** | **3011** | **1000** | **0** |
| **Trincomalee** | **5737** | **4668** | **142** |
| Total | 50,783 | 31,753 | 13,092 |

*Source:* RADA (2006)

critique on the role of INGOs and NGOs in Sri Lanka (Frerks & Klem 2005; *The Sunday Times* 2005; Amarasinghe 2007). Thus, the tsunami did not only place the capabilities and willingness of national governments in handling disaster and crisis under pressure, but also altered the missions and practices of development and donor organisations.

In the immediate weeks following the disaster of December 2004, international and local non-governmental organisations (INGOs and NGOs) pledged to rescue the affected coastal inhabitants from their suffering (relief), and donor agencies negotiated for collaboration with the government for making strategies to place the affected communities on better development paths by providing sustainable livelihoods and safer infrastructure facilities (recovery). There was competition between agencies and with so many actors involved coordination was challenging and may be said to have failed (cf. the failed P-TOMS initiative). However, in the implementation of housing construction there was a clear division of labour between the government and the INGOs: the government provided land, infrastructure and beneficiary lists; the INGOs funded and implemented the housing projects. As will be shown, the INGOs sometimes took intermediary positions between the government and the people affected by the tsunami to facilitate access to land and services (Brun & Lund 2008). The almost unlimited access to international funding in the months following the tsunami resulted in many INGOs becoming more powerful than before. They suddenly became significant donors in terms of their size and funds at their disposal. Some donors even 'disturbed' the traditional power balance between donors due to large inflows of money and other types of support. For example, The Red Cross Movement (and several other INGOs) became much larger in terms of personnel and cash flows than the UNDP, which traditionally had played a pivotal role in the donor environment in the country.

Not only did this situation disturb the power relations amongst donors, but it also changed the relationship between the Government and the international community in Sri Lanka. Authorities and donors had different mandates, needs and objectives. The GoSL emphasised the need for MMM (men, money, materials). Donors, however, identified other needs, had other abilities, or faced external demands. Most NGOs, for example, had to comply with external fund raisers who wanted to see results, often in the form of houses (even though it was not in their official mandate to build houses). Such conditionalities in fact led to inflexibility among donors locally, and their level and types of operations were curbed. Furthermore, approaches, knowledge levels and methodologies differed. Some agencies lacked the appropriate knowledge and competency for the work they were doing. As a result, the Tsunami Evaluation Coalition (TEC), for example, advised that NGOs should be accredited in order to avoid unnecessary incompetencies. TEC criticised the whole relief apparatus (Cosgrave 2007). As a result, incompetency and competition among donors continued to exist, and the government and public domain in general became more critical and even hostile towards the international community. This affected housing reconstruction in different ways, both directly and indirectly. More importantly, the debate on the role and status of INGOs has weakened their role in conflict and relief situations in Sri Lanka. However, this is not all due to their own doing. Unfulfilled policy goals of the GoSL have made the INGOs easy targets as scapegoats, and INGOs and other donors do not have any independence in the recovery process.

Furthermore, when the international actors started operating in Sri Lanka after the tsunami, they brought with them their own housing policy narrative: a narrative based on the international disasters discourse, with its own sets of understandings of recovery and of what housing should contain. Slogans such as 'building back better' refer to past practices and lessons learnt, with an emphasis on the importance of citizen's response to disasters within the wider contexts of building sustainable communities and not recreating nor exacerbating vulnerabilities. However, much of the post-tsunami 'building back better' documentation did not refer this prior work, and thus lost some of its potential to instigate development. Kennedy et al. (2008) show that several organisations have tried to link their tsunami emergency response budgets with their ongoing development budget, but have failed in their attempts. Often mandates did not permit such attempts, or the organisations did not have the capacity to make the link.

Another important dimension of the international community's understanding of housing was that of the material and physical structure of settlements. On visiting various planning offices along the coast of Sri Lanka, different housing models developed by different organisations could be found displayed on notice boards: the 'Care house', 'Oxfam house', etc. However, although different organisations were involved in housing reconstruction, standard housing models were followed. They were all based on a blueprint model with standardised costs and practicalities and not much developed in cooperation with the people in this particular area, grossly negating different understandings of what a house should be like, according to cultural practices and traditions. Building the good looking house – the label attached to an organisation's house a status of 'the good looking house' – seems to have been important. However, whilst there were standardised labels that ignored culture, the main challenges to emerge concerned how people would inhabit and use such houses. These models overlook the traditional diversity in settlement patterns and housing in Sri Lanka, and how traditional divides of class, caste and ethnicity are embedded in the design of settlements and housing. Another important element is the fact that the process of implementation was very much the same for all the donors: contractor based, no participation, or very little sub-contracting. They also overlooked the traditional Sri Lankan strategies of providing temporary houses and inputs for self-help and communal participation, such as is found in the Mahaweli project. Looking at the variety of houses built in the settlements for IDPs in Puttalam too, it is quite clear that people themselves – when they are given the chance – modify and personalise the layout of their houses.

The relationship between the Government of Sri Lanka and the international donor community also shifted during the post-tsunami implementation work. First, there was a shift in political leadership just after the tsunami occurred. While former president Chandrika Bandaranaike openly pledged foreign assistance, the new president, Mahinda Rajapakse, increasingly expressed scepticism towards the role of foreign donors in the country, especially Western ones. Hence, some INGOs collaborated with the authorities, while others did not. This situation negatively affected housing policies. Furthermore, criticism of biases, inefficiency and lack of coordination raised by the implementing agencies, such as the INGOs, was overlooked. At a later stage, this criticism was used as a reason for the GoSL to blame the INGOs for incompetence and deceit. This stemmed in part from not reaching the housing targets as planned and politically promised by the GoSL, and in part by the role of the INGO community in assisting victims – mostly Tamils – of renewed violence in the north and east. In a wider context, restricted donor abilities have led to questions regarding the reflexivity of the INGO community.

What may be deduced from this discussion is that the relationship between the international community of donors and the Sri Lankan authorities has not been static and it has changed from the first stage of relief work and need of incentives and money, to a stage of recovery, where the state has emerged as more decisive and less in need of external interventions. The issue of housing rests as a significant, but largely unresolved issue, however. Today, many of the INGOs have left the country and the implementation of services and maintenance has come to a halt. According to our own observations from the east, for example (September 2008), many houses too have been vacated, primarily due to lack of infrastructure such as water, proper roads, schools, and transportation, which should in effect be the responsibility of the Government.

## Land, beneficiaries and housing: multiple experiences embedded in the narrative

Government land for housing is distributed by the Divisional Secretariat (DS) and signifies the role of the mid-level or local-level government and civil servants in housing policies and practices in Sri Lanka. After the tsunami, these civil servants played a key role in the unfolding of events, but a key issue is also how they managed to collaborate with civil society and the international community.

### *Access to land*

The immediate response by the GoSL was to ban building in the areas closest to the sea, known as the 'buffer zone', in order to avoid another tsunami disaster. Prior to the tsunami the area adjacent to the coast was densely populated and was regulated by the Coast Conservation Act No. 57 passed in 1981. The implication of introducing the buffer zone, however, was that people's own recovery initiatives were put on hold. Instead of starting to rebuild their lives, they were waiting for new land, new housing, and new livelihoods to be provided. Although initially there was a general agreement on the need for a buffer zone by both the GoSL and the LTTE, the people who were directly and indirectly affected protested openly against what they saw as land grabbing on behalf of the authorities. The INGOs made similar protests, only more discretely. Local governments were squeezed between different demands from the national government, the international actors who wanted to start reconstruction work as soon as possible, and the people affected by the disaster that were also waiting to see the building of their future homes. Immediately after the tsunami, more than one-third of households who had lost land during the tsunami reportedly preferred to rebuild their houses in the same place as before (Institute of Policy Studies 2005).

However, after 11 months – just prior to the presidential elections – the buffer zone was changed – almost without public attention, although it affected the lives of 100,000 people and reduced the need for relocating housing.[5] Today, there is still lack of clarity of ownership in the buffer zone and beyond, and there is a need to make people aware that land 'ownership' in this regard covers those who live on state land. Flexibility in land allocation is thus essential, but according to the Institute of Policy Studies (2005) there has been no flexibility on this matter. Lack of sufficient and appropriate land led local officials to claim that it was not possible to follow national and international standards (Kennedy et al. 2008). The tsunami was interpreted as so exceptional that they just had to do whatever they could to find land and building houses as fast as possible.

One example from Ampara may serve to illustrate some of the key dilemmas and issues as they were experienced by the people. In the Eastern Province of Sri Lanka, one of the main affected areas, land was already a sensitive issue due to demographic shifts and territorial claims that had emerged during the war (Brun & Lund 2008). In the three Districts of the Eastern Province the buffer zone was set at 200 m. In the context of land scarcity and land disputes, the buffer zone created a number of challenges in terms of finding enough suitable land for all of the affected families that needed to relocate and build new homes after the tsunami, and, not least, a high degree of insecurity and fear for their futures were felt by such families. The Government's reduction of the buffer zone in 2006 created further confusion on the part of locally affected people, but also showed how the position of already marginalised groups of the population were reinforced. In Oluvil village, for example, a group of mainly fishing families, who had all lived within the 200 m buffer zone established after the tsunami, and all of whose houses had been damaged, were granted houses in a jungle area 6 km distant from the sea. However, when the buffer zone changed, almost half of the families were allowed to occupy their old plots. Confusion arose among the people who were told by the local authorities that if they did not accept the assistance offered, i.e. the new houses granted in the jungle, they would not necessarily be entitled to assistance to rebuild their damaged houses by the coast. In this situation, the people felt compelled to turn to an INGO to build their houses, and to assist in negotiations with the government and give advice.

Furthermore, entitlements have not been handled in the same way across all housing and settlement schemes. In the Mahaweli settlements, a land title is called *Parithyage oppuwa* (gift). This cannot be sold, mortgaged or uneconomically divided, but it can be handed over to the elder son or to a widow. In some other settlements initiated by the government, as well as in the post-tsunami settlements on public land, the land provided to the settlers is given either on a deed (*swarnabumi*, introduced in 1982), or on a permanent lease (99 years), which requires a yearly payment (*balapatrayak*). Leased lands may be sold or mortgaged to obtain loans, with the permission of land commissioner. In the tsunami settlements, some people have been given land deeds while others have not yet received them (fieldwork notes 2008).

### *Modes of participation*

After the tsunami, there was limited advocacy for a people-centred reconstruction process. Because rebuilding as fast as possible became the aim, reconstruction practices concentrated on constructing buildings, roads and infrastructure, but overlooked rebuilding communities, community lives and livelihoods. Thus, a huge number of displaced people suddenly found themselves in absolute poverty and destitution, and needing to 'regain their lives'. In addition, small and dispersed settlements, whose inhabitants were drawn from different villages, did not have either the magnitude in terms of numbers or social cohesion to create new income generating activities. Part of the construction work was based on self-help and community participation, even though (as already mentioned) in the newly formed settlements a donor-driven approach dominated for the first two years. Furthermore, an effectively cooperating 'community' entails a high level of social trust, hence community rebuilding requires the participation of people who trust or might have a perceived reason to trust each other. This makes selecting 'beneficiaries' a fundamental issue. After the tsunami this was grossly underestimated by the District Secretary's offices, which were delegated the task of deciding who were to benefit and where, from new relocation housing schemes. Therefore, a number of developments occurred as unplanned, uncoordinated consequences of assistance strategies. The coping strategies applied by people themselves were basically overlooked, as were the needs of vulnerable and marginalised groups (the elderly, children, women, disabled). Instead, a plethora of arbitrary decisions emerged. Different donors responded differently to key issues such as housing design, community, and people's needs, gender and ethnic differences, and articulated housing needs.

Visits to sites in Hambantota District in the three years after the tsunami revealed some interesting facts on who are the real beneficiaries of these housing schemes: the surplus of 3645 (more than 5000 according to more recent estimates) tsunami housing units shows that there are bound to be numerous 'beneficiaries' that hold dubious claims to the right of a free, new house. Demonstrably, there are poor people who have been awarded houses that will fundamentally change their lives for the better. Amongst those who live in these schemes are a substantial number of house owners without legitimate 'tsunami claims' (see also Perera 2007).

### *Build back faster?*

From both victims and the GoSL – and the national and international press – we have witnessed considerable criticism of lack of speed in the housing reconstruction process. The implementing agencies, most of which were INGOs, were brought into housing construction for the very first time (Lions Club, The Red Cross movement, PLAN, Medecins Sans Frontiers, and numerous new local NGOs). Both these and those more experienced either stalled or were held back due to lack of available building sites, through not being given adequate 'beneficiaries lists', as a consequence of incoherent mandates from the GoSL, and through a general reluctance to comply with being relegated to the role of mere 'building managers'. At the same time, these organisations were pressed by their constituency at home to deliver. As a consequence, we saw an overwhelming activity in the south where the GoSL had prepared for massive reconstruction projects. In Hambantota, the President's home constituency, approximately five times the number of housing units have been built – primarily by INGOs – as those destroyed (Table 1). This allowed organisations to report back that thousands of houses were being built in Sri Lanka with the money given by people in donor countries.

Rebuilding houses quickly may not be the best strategy when thinking in terms of social and economic recovery and long-term development. The post-tsunami situation did not allow answering questions such as how are new places, villages and communities planned, built and inhabited, and by whom, and who is to select 'worthy beneficiaries', and on what grounds. The aims of rebuilding houses as fast as possible resulted in (lack of) understanding of strategic place making and site development in the reconstruction stage (see Brun & Lund 2008, on the acceptable time dimension for rebuilding and constructing houses after the tsunami). In this way, the post-tsunami reconstruction differed substantially from that of the Mahaweli project, where the time span was 6–30 years, and when we visited in 2004, several of the settlers had still not managed to build a new house and still lived in modifications of the initial core mud house granted to them upon resettlement, which indicates poor performance in the scheme and lack of well-being.

## Challenging the narrative

The Indian Ocean tsunami of 2004 has been considered as exceptional. Due to its exceptionality, the recovery process did not take on board experiences from past housing projects. We find patterns, similarities and differences in the cases presented here that can assist in understanding the policy processes and the policy narrative. While much of the same language and practices follow through the cases presented, it is not easy to find evidence for a uniform One National Housing Policy in Sri Lanka. Thus, the narrative has to be contextualised with respect to practices, stakeholders, processes, and 'language' involved at different scales.

In their various contexts, the dimensions of housing policies provide insights into lessons learnt and for analysing some of the underlying motivations and aims of polices. The most common language of all cases is perhaps the unspoken aims of the housing policies and practices – aims that may not always be about housing for people affected by poverty, war or disaster, indicating that housing is often a tool for other strategic political aims. Rather than ONHP, the politics of land and identities may be seen as inherent in the cases described above. Since the Mahaweli project was instigated, conflict and ethnic segregation have both influenced and emerged as a result of housing policies and plans. While control over land, resources and power is the dominant reason for this, it is also interesting that little or no attention has been given to *housing diversity* in Sri Lanka. Different groups have different housing needs, as made manifest in pre-tsunami land use patterns and housing and settlement plans. In all cases, from the Mahaweli project to the tsunami, we have seen that settlements are organised, not according to the traditional settlement patterns, but rather as blueprint models of nucleated settlement patterns, leading to densely populated housing units. This has become the tradition of organising settlements in Sri Lanka today. Along with this highly centralised settlement structure and top-down planning and implementation, the close links between settlement leaders and the authorities show how politics are embedded in the everyday lives of people in Sri Lanka.

This leads us to the role of politics in land distribution and housing, which seems to have led to a tendency of location being more about settling people in politically strategic locations rather than in relation to livelihoods and other basic issues for people's lives. However, it is not only about high-level politics. Land is also related to security and feelings of home and identity. In all cases presented, most people's acceptance of policies has been motivated by the need to make a new home and life in an area that represented various forms of (in)security. However, as we have also seen in all cases, issues of property deeds, people's status, and the lack of long-term thinking related to land size, have caused a number of challenges for livelihood security and for future generations. In this regard, the various types of title one can have to land in Sri Lanka has meant that new housing settlements often tend to limit people's long-term security. Consequently, insecurity is strengthened by the fact that, in all cases studied people had moved away from their former home. Displacement meant that the social fabric and former livelihoods became disturbed.

In all cases, a joint language of land and housing based on one size and one set of immediate needs prevails. This policy applied to housing reconstruction after the tsunami was exacerbated by the contextual ignorance of the incoming INGOs and the expediency – and ethnic bias – of the GoSL. In addition, standard plans were developed and land use patterns of new settlements were grounded in what could be labelled a 'Singhalese, urban middle class' housing pattern. The affected people in the east, mainly Muslims and Tamils, had other priorities as, evidently, did the rural Singhalese of the south. Despite these realities, the incoming INGOs all abided by the call for standard housing solutions in spite of different cultural housing and land use patterns. This policy is, in fact, not new to Sri Lanka. In all big development projects, such as the Mahaweli project, housing for the poor and landless has been an intrinsic project component. Some of these development projects have also been criticised for their Singhalese bias, as much land and housing has been allocated to landless and poor Singhalese people.

While we cannot find evidence for one national housing policy, an interesting feature is the continued importance of the government throughout all cases. In Sri Lanka, housing is a key political symbol for successful political measures. Housing is very visible and a crucial dimension to people's lives, and is thus used to mobilise political support. Additionally, housing is clearly used as a way of controlling demography, ethnic composition, and the mobility and location of people. Perhaps as a consequence of the dominant language of housing in the Sri Lanka polity, housing has come to mean not only the right to shelter, but also the right to house ownership.

In the Mahaweli and Puttalam cases, we have seen a gradual approach to housing. The housing process would typically start with a temporary hut, then the foundations of a permanent house would be built, and in time the house would eventually be completed. In this way, people themselves gradually built their own houses and played a role in designing and developing their homes. While poverty, and also war to a certain degree, are considered catalysts of change, disasters too can be agents of social and political change. However, a disaster is often understood as an exceptional moment in time that requires exceptional methodologies. After the 2004 tsunami disaster, it was claimed that housing had to develop quickly, people were not to be burdened with building their own houses, and there was no time to involve people in the process. Compensation came to mean being given a ready-made house. Organisations that would normally apply participatory methods were pushed into, and even themselves advocated non-participatory methods in the construction of houses after the tsunami.

Another crucial difference between the Mahaweli, the Puttalam, and the post-tsunami cases is the presence of international organisations in housing construction. Although international funding played a fundamental role in the Mahaweli project, the kind of housing received little attention from international donors, and construction of housing was, to a large extent, instigated by the settlers. In Puttalam, people organised themselves to build houses, acquire land and organise their settlements with little outside assistance. The presence of international non-governmental organisations was very limited in the case of housing in Mahaweli and to some extent in Puttalam. In light of this experience, it is interesting to see the shift after the tsunami. In the Sri Lankan case, and unlike in India and Thailand, which both took a different approach to the reconstruction (*Forced Migration Review* 2005), there was an unprecedented presence of international organisations playing a role in the housing programmes.

Finally, there are reasons to ask what difference the international community made to housing policies after the tsunami. Criticism of biases, inefficiency and lack of coordination raised by both the Government and foreign donors reflected lack of mutual partnership arrangements and mandates. It could also be asked how the INGO system was deprived of the chance to effectively collaborate with the Government and how significant the political situation and the ONHP narrative were in this regard.

## Conclusions

This article has asked whether housing is a meaningful way to 'do' development. However, there is no one way to develop through housing. It is found that the dominant national housing policy narrative has made multiple knowledge and past experiences on housing invisible. Although some past experiences are embedded in new settlement schemes, it is not necessarily the 'best practices' that are carried into the new situations. Besides, too much emphasis has been put on physical reconstruction and too little on how people have been enabled to participate in developing houses and livelihoods.

We have shown that it should have been possible to learn from past experiences: various studies have shown how the changes from small, homogeneous family compounds to large, densely populated and heterogeneous settlements in Mahaweli brought about massive socio-economic and social change (Lund 1981; de Soyza 1995; Schrijvers 1986; Raby, cited in Mathur & Cernea 1995). Likewise, Puttalam is composed of different groups, i.e. settlers, donors and officials pursue their often divergent interests. The degree of freedom available for the different groups in both areas is determined by a number of systemic constraints related to the size and multifunctionality of the housing and settlement programmes, entitlement to land and the speed of construction (Mueller & Hettige 1995). Re-examining such experiences enables us to learn something about the successes and failures of the housing processes and which are the key issues and bottlenecks in planning and implementing resettlements and housing.

Tensions in the settlement process have taken place also throughout the planning and implementation stages of the programmes and up to the present, related to competition over resources, bureaucratic control and top-down management. Added to technocratic bias in implementation and lack of coordination, most housing programmes have been ridden with attitudinal problems and also ethnic and class biases. Furthermore, procedural issues that need to be clarified prior to any resettlement and reconstruction effort, including the post-tsunami one are: access to titles to property and land, an unbiased selection of beneficiaries, appropriateness of the speed of implementation, and the role of and coordination among external implementing and participating actors and agencies.

Despite appropriate slogans and good intentions, the question remains: Was the situation after the tsunami so unique that it necessitated deviation from previous policies and practices – and if so, how? Although the planning model seems to be similar in all cases – that of a highly centralised and top-down manner of execution (which also forms the basis for the narrative itself) and a standardisation of land size, benefits and services – post-tsunami projects have been different from other schemes in that the houses constructed were of a much higher standard, their implementation was different due to service provision by many different internal and external actors, all with different agendas. Their ignorance of past experiences and inability to understand current policies and power relations have led to a silent acceptance of ONHP.

This, in turn, led to limited participation of the primary stakeholders – the tsunami affected people themselves – who were not sufficiently integrated in the implementation of post-tsunami housing projects. In this regard, very little was 'learnt' from previous housing and settlement schemes, where people's own participation was a key element. The housing policy after the tsunami thus stands as unique in that it did not sufficiently embed its practices in local realities and people's own preferences and possible contributions. Future national housing policies in Sri Lanka should take this learning into account.

## Notes

1 Equal authorship.

2 It is common to name 'settlers' those who came to the resettlement schemes. Displaced persons have become the common term for the conflict-induced displacees. During the 1990s, with increasing emphasis on displacement both by the international community and national actors, the term displaced people is used on a wider set of people moving under different degrees of force within Sri Lanka.

3 R. Lund 'What can we learn from previous Sri Lankan experiences? The case of the Accelerated Mahaweli Programme'. NTNU/FORUT international workshop on housing, 26–27 May 2005, Colombo. Unpublished.

4 Recently (during 2007) the World Bank started a new initiative in Puttalam to relocate IDPs. This process has not been included in the discussion presented here.

5 With respect to the ideas behind the buffer zone as a policy and how it affected the housing strategies after the tsunami, we refer to our Ampara example presented in the following discussion, and also Hyndman's article (2009, this Issue).

## References

Amarasinghe, S. 2007 *'I/NGOs: Mea Culpa...Your Culpa...or Our Culpa?'* Serving Sri Lanka. http://servesrilanka.blogspot.com/2007/12/ingos-mea-culpayour-culpaor-our-culpa.html (accessed 10 January 2008).

Bastian, S. 1996. Control of state land – the devolution debate. ICES. (ed.) *Sri Lanka: The Devolution Debate*, 61–86. International Centre for Ethnic Studies (ICES), Colombo.

Bennett, J., Bertrand, W., Harkim, C., Samarasinghe, S. & Wickramatilake, H. 2006. *Coordination of International Humanitarian Assistance in Tsunami-affected Countries.* Tsunami Evaluation Coalition (TEC), London.

Blaikie, P.M., Springate-Baginski, O., Banerjee, A., Bhatta, B., Saigal, S. & Sarin, M. 2007. Actors and their narratives in participatory forest management. Springate-Baginski, O. & Blaikie, P.M. (eds.) *Forests, People and Power: The Political Ecology of Reform in South Asia*, 92–115. Earthscan, London.

Braunmuhl, C. von Boltz, R. Jayatilake, L. Noble, K. & Saroor, S. 2006. *Tsunami: A Study on Disaster in Sri Lanka*. Report prepared by Bread of the World, Stuttgart, Medico International, Frankfurt/Main and the Heinrich Boll Foundation, Berlin. http://www.medico-international.de/material/downloads/mi_tsunami-report_2006_it.pdf (accessed 10 December 2007).

Brun, C. 2008. *Finding a Place. Local Integration and Protracted Displacement in Sri Lanka*. Social Scientists' Association, Colombo.

Brun, C. & Lund, R. 2008. Making home during crisis: Post-tsunami recovery in a context of war, Sri Lanka. *Singapore Journal of Tropical Geography* 29:3, 274–288. DOI: 10.1111/j.1467-9493.2008.00334.x.

Clinton, W.J. 2006. Key propositions for building back better. Lessons learnt from tsunami recovery. http://www.tsunamispecialenvoy.org/pdf/SELLR_122506.pdf (accessed 15 March 2007).

Cornwall, A., Harrison, E. & Whitehead, A. 2007. Gender myths and feminist fables: The struggle for interpretive power in gender and development. *Development and Change* 38:1, 1–20.

Cosgrave, J. 2007. *Synthesis Report: Expanded Summary. Joint Evaluation of the International Response to the Indian Ocean Tsunami*. Tsunami Evaluation Coalition (TEC), London.

David, A.K. 1986. Mahaweli: 'What Next? *Sri Lanka Journal of Social Sciences* 9, 1–2.

de Soyza, D.A. 1995. *The Great Sandy River. Class and Gender Transformation among Pioneer Settlers in Sri Lanka's Frontier*. PhD thesis, University of Amsterdam, Amsterdam.

Fernando, U. & Hilhorst, D. 2006. Everyday practices of humanitarian aid: Tsunami response in Sri Lanka. *Development in Practice* 16:3, 292–302.

*Forced Migration Review*. 2005. Tsunami: learning from the humanitarian response. *Forced Migration Review*. Special Issue July 2005.

Frerks, G. & Klem, B. 2005. *Tsunami Response in Sri Lanka*. Clingendael Institute, The Hague. http://www.clingendael.nl/publications/2005/20050300_cru_other_frerks.pdf (accessed 12 December 2007).

Hasbullah, S.H. 2001. *Muslim Refugees. The Forgotten People in Sri Lanka's Ethnic Conflict*. Research and Action Forum for Social Development, Nuraicholai.

Hyndman, J. 2007. The securitization of fear in post-tsunami Sri Lanka. *Annals of the Association of American Geographers* 97:2, 361–372.

Hyndman, J. 2009. Siting conflict and peace in post-tsunami Sri Lanka and Aceh, Indonesia. *Norsk Geografisk Tidsskrift–Norwegian Journal of Geography* 63:1, 90–96.

Government of Sri Lanka . 2005. *Post-Tsunami Recovery and Reconstruction*. A joint report of the Government of Sri Lanka and its development partners. http://www.tafren.gov.lk/portal/index.jsp?sid = 1&id = 76&pid = 1 (accessed 15 March 2006).

Institute of Policy Studies. 2005. *Listening to Those Who Lost: Survey and Analysis of Rebuilding and Relocation of Tsunami Affected Households in Sri Lanka*. Institute of Policy Studies, Colombo.

Karunatilake, H.N.S. 1989. Some significant economic aspects of the Accelerated Mahaweli Programme. *Sri Lanka Economic Journal* 4:2, 1–11.

Kennedy, J., Ashmore, J., Babister, E. & Kelman, I. 2008. The meaning of build back better: Evidence from post-tsunami Aceh and Sri Lanka. *Journal of Contingencies and Crisis Management* 16:1, 24–36.

Lang, H. & Knudsen, A. 2009. 'Your subject of protection is a dangerous one': Protracted internal conflict and the challenges for humanitarian agencies. *Norsk Geografisk Tidsskrift–Norwegian Journal of Geography* 63:1, 35–45.

Lewis, J. 1981. The Sri Lanka cyclone 1978: Socio-economic analysis of housing destruction. *Marga* 6:2, 1–33.

Lewis, J. 1999. Vulnerability to a cyclone: Damage distribution in Sri Lanka. Case Study IV. IT Publications (ed.) *Development in Disaster-prone Places: Studies of Vulnerability*, 87–108. IT Publications (Practical Action). London.

Long, N. 2001. *Development Sociology. Actor Perspective*. Routledge, London and New York.

Lund, R. 1981. Women and development planning in Sri Lanka. *Geographical Annales* 63B, 2.

Lund, R. 1983. The need for monitoring and result evaluation in a development project – Experiences from the Mahaweli project. *Norwegian Journal of Geography* 37, 176–186.

Lund, R. 1993a. *Gender and Place. Towards a Geography Sensitive to Gender, Place and Social Change*. Vol. 1. Department of Geography NTNU, Trondheim.

Lund, R. 1993b. *Gender and Place. Towards a Geography Sensitive to Gender, Place and Social Change*. Vol. 2. Department of Geography NTNU, Trondheim.

Mathur, H.M. & Cernea, M.M. (eds.) 1995. *Development, Displacement and Resettlement. Focus on the Asia Experience*. Vikas, New Delhi.

Mueller, H.P. & Hettige, S.T. (eds.) 1995. *The Blurring of a Vision – The Mahaweli. Sarvodaya*. Book Publishing Services, Ratmalana.

Perera, S. 2007. *Sri Lanka: Not All Tsunami Reconstruction is Equal*. IPS, http://ipsnews.net/news.asp?idnews = 38676 (accessed 2 October 2008).

RADA. 2006. Development assistance data base. UNDP and RADA, Colombo. http://www.rada.gov.lk/portal/resources/_housing/SF%20Present%20status%20housing%20Nov%2003.xcl.pdf (accessed 24 October 2005).

Saravananthan, Mutthukrishna 2007. *People's Verdict on Tsunami Recovery in Sri Lanka*. International Centre for Ethnic Studies, Colombo.

Schrijvers, J. 1986. *Mothers for Life. Motherhood and Marginalization in the North Central Province of Sri Lanka*. Eburon, Delft.

Scudder, T. 2005. *The Future of Large Dams. Dealing with Social, Environmental, Institutional and Political Costs*. Earthscan, London.

Silva, K.T. 2009. 'Tsunami third wave' and the politics of disaster management in Sri Lanka. *Norsk Geografisk Tidsskrift–Norwegian Journal of Geography* 63:1, 61–72.

Skotte, H. 2004. *Tents in Concrete. What Internationally Funded Housing Does to Support Recovery in Areas Affected by War: The Case of Bosnia-Herzegovina*. PhD thesis, NTNU, Trondheim.

Steele, P. (ed.) 2006. *Livelihoods in Post-Tsunami Sri Lanka: 'Building Back Better'?* Institute of Policy Studies and International Labour Organisation, Colombo.

Sørensen, B.R. 1996. *Relocated Lives: Displacement and Resettlement within the Mahaweli Project, Sri Lanka*. Sri Lanka Studies 3VU. University Press, Amsterdam.

Tete, S. 2009. Whose solution? Policy imperatives vis-à-vis internally displaced persons' perceptions of solutions to their situation in the Sri Lankan conflict. *Norsk Geografisk Tidsskrift–Norwegian Journal of Geography* 63:1, 46–60.

*The Sunday Times*. 2005. Abuse of tsunami funds by NGOs. Tyron Devotta. 3 April 2005. Wijaya Newspapers Ltd., Colombo.

Uyangoda, J. 2005. Ethnic conflict, the state and the tsunami disaster in Sri Lanka. *Inter-Asia Cultural Studies* 6:3, 341–352.

# Livelihoods in a conflict setting

JOHN NIGEL

The ethnic conflict in Sri Lanka is a major threat to livelihoods of all people, especially those in rural areas. The study aims to identify and discuss the impact of ethnic conflict on the livelihoods of households in the conflict-affected areas, based on the experiences of those living in Ampara District. In the study area, conflict-related violence is not only limited to Sinhalese and Tamils, but also there are sporadic clashes among Tamil militant groups themselves, and between Tamil and Muslim communities. Conflict has had a devastating effect on the livelihoods of people in the conflict areas, resulting in widespread hardship. Research reveals that people create strategies to diversify their livelihoods by using different assets that they are able to access in conflict situations.

*John Nigel, Department of Geography, Norwegian University of Science and Technology*

## Introduction

Since 1956 Sri Lanka has experienced several incidents of ethnic violence. The worst of these erupted in 1983, and rapidly escalated into a civil war between the Liberation Tigers of Tamil Eelam (LTTE) and the government of Sri Lanka. The LTTE claims an independent Tamil homeland in the northern and eastern parts of Sri Lanka and the state deploys large numbers of troops in its fight against the LTTE to safeguard the territorial integrity of Sri Lanka. The fighting became more intense from the 1990s, with devastating effect on the livelihoods of the population. Although the whole of the island is affected by war-induced insecurity, the Northern and Eastern Provinces have been the main areas of heavy fighting in the past quarter of a century of war. These two provinces are predominantly occupied by Tamils, followed by Muslims, while remaining part of the country is predominantly occupied by Sinhalese.

The 25-year war has claimed more than 70,000 lives (an average of eight lives per day), and resulted in more than 100,000 physically harmed, almost 1 million displaced, and the destruction of more than 500,000 private and public properties (Human Rights Watch 2007). Hopes of lasting peace were raised when the Sri Lankan Government and the LTTE signed a Norwegian-mediated ceasefire agreement and a Memorandum of Understanding (MOU) in February 2002. Attempts were made to restore the lives of the people, rebuild the war-torn areas, find a political solution to the war and establish ethnic harmony and peace in the country, but as stated by Bohle & Fünfgeld (2007), the ceasefire was fragile from its inception, and it was aborted in January 2008.

The persistent uncertainty and feeling of insecurity has been increased by the impacts of the tsunami of 26 December 2004. It was a double blow to the communities living near the coast who had undergone severe hardships during a quarter of a century of civil conflict. The tsunami struck along two-thirds of Sri Lanka's coastline. The damage stretched from Jaffna in the north, along the entire eastern and southern coast, and extended to north of Colombo on the west coast. The tsunami affected more than 1 million people, killed 35,322, and destroyed 150,000 livelihoods (PTRR 2005). Thus, it reasonable to assume that war-affected communities have been more vulnerable to the combined effects of the tsunami of 2004 and war compared to those which have affected by war alone.

The background to Sri Lanka's ethnic conflict has been well documented (de Silva 1984; Bastian 1985; Hellmann-Rajanayagam 1990; Shastri 1990; Fernando 1997; Rogers et al. 1998; Gamage & Watson 1999; Seneviratne 1999; Silva 1999; Snodgrass 1999; Chandrakanthan 2000; Wilson 2000; Korf et al. 2001; Brun 2003a; Jayawardena 2003; Abeyratne 2004; Hasbullah et al. 2005; Uyangoda 2005; Brun 2008). There have also been a number of studies within development thought linking conflict and livelihoods (Goodhand et al. 2000; Jaspers & Shoham 2002; Schafer 2002; Bhatia et al. 2003; Collinson 2003; Korf & Silva 2003; Silva 2003; Korf 2004; Stites et al. 2005; Lautze & Roberts 2006; Ruwanpura 2006; Bohle & Fünfgeld 2007; Lautze 2008; Shanmugaratnam 2008). Nonetheless, those studies which primarily focus on the nexus between ethnic conflict and livelihood in the war zones in Sri Lanka are rare. Of the few studies that have explored the situation in the north and east, most have focused on the impact of displacement process and the situations of internally displaced persons (Brun 2003b; Sidharthan 2003; Silva 2003; Muggah 2008). The present study will examine how livelihoods have changed over time in rural Ampara District during the conflict and also the differences in livelihood strategies among the different ethnic groups.

Goodhand et al. (2000) have explored the links between militarized violence and social capital formation in the Batticaloa District in the Eastern Province. Based on case studies, they identified that the links between militarized violence and social capital are complex, dynamic and context specific. They found that the political economy of war has created social capital within some people's own ethnic group but destroyed it between ethnic groups.

Korf et al. (2001; 2004) carried a study of the livelihood strategies of war-affected communities in Trincomalee District (Eastern Province), by considering all three communities (Tamil, Sinhalese and Muslim). He argues that livelihood strategies and the outcomes of the different communities differed considerably depending on very specific local political geographies despite the fact that they lived close together geographically. He concludes that war can be both a threat and an opportunity; hence, civilians in conflict situations are not all victims. Korf & Fünfgeld (2006) have investigated how the political geographies of war affect access regimes and entitlements to common-pool resources and thereby confine the livelihood opportunities of resource users in peasant and fishing villages in the war-affected Batticaloa and Trincomalee Districts. Their study discusses the way in which different peasant and fishing groups used various different opportunities which emerged in their political field at a specific time and place for small opportunity gains at the expense of other ethnic categories. Bohle & Fünfgeld (2007) have carried out a case study focusing on the political ecology of violence in the Batticaloa District of eastern Sri Lanka. They found that livelihoods became highly politicized and how the politicization of resource-based livelihoods created alternative systems of power and control over natural resources. Shanmugaratnam (2008) has investigated the dynamic processes of livelihoods of Tamils in predominantly agrarian localities of Batticaloa District. The study mainly concentrated on people's perceptions of their livelihood status and the changes they had experienced over time.

Complementing the existing livelihood studies in the war-affected Eastern Province of Sri Lanka, the present study will investigate the empirical realities of the livelihood situation under conditions of conflict using qualitative research methodology by examining the rural villages in Ampara District in eastern Sri Lanka. This study is the first to document the livelihood experiences of the people representing all three major ethnic communities in the district, making this study unique within the livelihood literature on the conflict in Sri Lanka.

## The conceptual framework of livelihoods

Livelihood studies were carried out by scholars from different disciplines in various themes such as risk and hazards relating to ecological disaster (Blaikie et al. 1994; Gilbert 2001; Wisner et al. 2004); problems of sustainable livelihoods (Chambers & Conway 1992; Carney 1998); diversification (Reardon et al. 1992; Ellis 2000); and power and conflict (Jaspers & Shoham 2002; Collinson 2003; De Waal, 2005; Lautze & Roberts 2006; Lautze 2007). All of these studies include a focus on the strategies and practices pursued by people in their day-to-day struggle to making a living.

Livelihood approaches have increasingly playing a key role in development contexts in recent years. In the policy debate, the first appearance of livelihood approaches can be traced in the Brundtland Commission Report of 1987 (Solesbury 2003). In the early 1990s, Chambers & Conway (1992) helped to consolidate the concept by proposing a definition for sustainable livelihoods. Subsequently, Carney (1998) presented a definition, based on Chambers & Conway's definition, which consolidated a number of different livelihoods frameworks currently in use:

> A livelihood comprises the capabilities, assets (including both material and social resources) and activities required for a means of living. A livelihood is sustainable when it can cope with and recover from stresses and shocks and maintain or enhance its capabilities and assets both now and in the future, while not undermining the natural resource base. (Carney 1998, 4)

Carney's definition also refers to the fact that livelihoods become vulnerable when unsuitable strategies undermine the natural resource base.

From the early 2000s, modified versions of the definition of sustained livelihoods have been generally adopted. According to Ellis (2000), a livelihood encompasses income, both in cash and kind, as well as the social institutions (kin, family, compound, village, and so forth), gender relations, and property rights required to support and sustain a given standard of living. He further describes the commonly accepted core of sustainable livelihood thinking as the requirement to understand and act upon the asset limitations of the poor, the risks they confront, and the institutional environment.

The livelihood approach used here is based on the idea that changes in livelihood strategy are normally adaptive responses to changes in six key interacting assets components:

(1) Natural assets – these cover natural resource stocks such as agricultural land, water resources, forests, environmental resources, etc., from which resources useful to livelihoods are derived.
(2) Social assets – social resources such as networks, kinship structures, relationships of trust, neighbourhoods, religious groups, etc., upon which people draw in pursuit of livelihoods.
(3) Political assets –include access to political power holders and citizenship status, necessary to pursue different livelihood strategies.
(4) Physical assets – basic infrastructure such as agricultural equipment, sewing machines, transport, livestock, shelter, etc.
(5) Financial assets – financial resources such as wages, access to credit, savings, regular remittances, etc.
(6) Human assets – these cover skills, knowledge, health, psychological reactions, etc., which are important to the ability to pursue different livelihood strategies.

The concept of livelihood strategy has become central to development practice in recent years and is particularly useful in conflict situations where chronic factors of increasing security risk, degradation of economic activities, and escalation of violence play a key role in shaping people's livelihood strategies. Hence, a clear understanding of how people pursue their livelihood strategies is of paramount important in determining the behaviour of households in the conflict areas.

The strategies of livelihoods mainly refer to the ways in which people combine their income generating activities, the way in which they use their assets, how they are forced to reorganize or cut back on their levels of consumption, and how they manage to preserve existing assets and income. Households may find themselves either struggling to survive in the most severe situations, or adopting coping strategies (short-term) or adapting strategies (long-term) to deal with sudden shocks or gradual trends (Rennie & Singh 1996).

According to Jaspers & Shoham (2002), in conflict situations households respond by adapting their strategies to ensure the survival of the household. In such situations, people affected by conflict generally strategize for surviving immediate threats and overcoming long-term threats to their livelihoods. In cases of protracted conflict, people use mixed strategies of balancing immediate needs and long-term survival. As stated by Young & Osman (2006), livelihoods cannot be considered simply in terms of household access to resources and capital and related livelihood strategies. In fact, conflict situations, chronic factors of increasing security risk, degradation of economic activities, and escalation of violence all play a key role in shaping livelihood strategies across the range of assets available. Hence, a clear understanding of how people follow their livelihood strategies is of paramount importance in determining the behaviour of households in conflict situations.

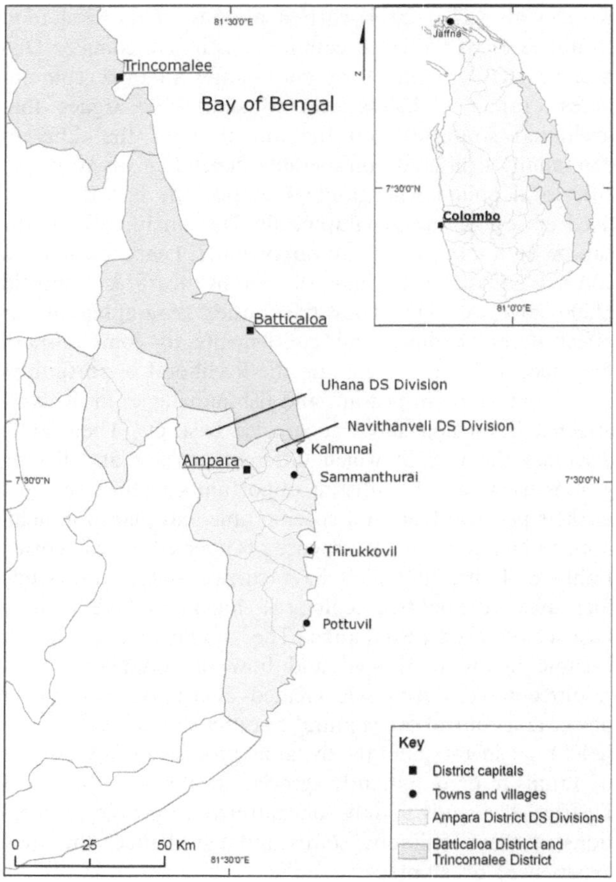


*Fig. 1.* Ampara District: location of the study area (shown shaded black on the inset map)

## Ampara District

Ampara District is located in the south-east of Sri Lanka (Fig. 1). It is one of three districts in the Eastern Province. This is a multi-ethnic, multi-religious, and multicultural district. The population of the district in 2007 was 610,719, of which 44% were Muslims, 37.5% Sinhalese, and 18.3% Tamils, and with 50.4% males and 49.6% female (DCS 2007). The Sinhalese largely live in the great majority of the interior area of the district, while the Tamil and Muslim populations are located mainly along the coastal belt, often in close proximity. Before the advent of armed conflict in 1990, there relations between the two latter communities were cordial, but conflict has created increasing mistrust and misunderstanding between them (Hasbullah et al. 2005).

Although Ampara District has experienced sporadic clashes among the Tamil militant groups, and also between Tamil and Muslim communities, the main conflict has been closely linked to the ethnic conflict between the LTTE and the Sri Lanka Army (SLA). The first major clashes between the LTTE and SLA originated in Trincomalee and spread to other areas in the east, including Ampara District in June 1990. From 1987, the LTTE was at war with the Indian Peace Keeping Force (IPKF). During the period 1987–1989, a large number of Tamils and Muslims underwent ruthless harassment by the IPKF and non-LTTE Tamil militants groups.

When the IPKF withdrew in March 1990, the LTTE took over control in major parts of the north and east and this was a time characterized by increasing repression by the LTTE against the Muslims in the east. This marked the end of Muslim support for the LTTE and the beginning of Muslims engaging in anti-Tamil violence (UTHR 1990; Brun 2003). In June 1990, the SLA moved into Ampara District. When the SLA advanced towards the main towns in the Tamil areas, the LTTE then moved into the jungle. This operation resulted in the displacement of large numbers of people, mainly Tamils and some Muslims. Throughout the 1990s the fighting between the two groups intensified. The LTTE gradually infiltrated the district and brought under its control a small portion of the remote rural area along the northern border and a large rural area in the southern part. The government forces had the control of the entire Sinhala area, urban and peri-urban settlements, and the major roads.

In 2002, military combat between the LTTE and SLA was suspended by the Memorandum of Understanding (MOU) and attempts were made to restore the lives of the people. In March 2004, the LTTE's eastern commander, known as Karuna, a powerful Tiger leader credited for many of the rebels' victories over security forces, broke away from the LTTE and to maintain a stronghold in the south-east where his fighters regularly attack the LTTE. The undoubted difficulty of negotiating for a win-win solution for both the government and the LTTE, and also the breakdown of peace talks between the LTTE and the leaderships of five different governments in the past, has convinced many in the government (and in larger society) and the LTTE that a negotiated settlement is simply not possible. This widely held view may possibly have resulted in the resumption of military operations in mid-2006. In July 2007, the government declared that all of the Eastern Province had been brought under its control. It was widely believed that the Karuna group backed the SLA during the operation, although the government has strongly denied this.

## Fieldwork methodology

The fieldwork was carried out during October 2007–April 2008 in three agriculture-based villages in the Ampara District of eastern Sri Lanka, using qualitative methodology. The district was selected mainly for four reasons. First, this is one of the districts which were directly affected by the conflict. Second, it represents all three major ethnic communities. Third, this is a less tense area in the conflict zone, allowing sufficient confidence to conduct research within acceptable levels of risk. Fourth, and finally, none of the academic research conducted in this district has explored the impact of conflict on livelihoods.

One of the major constraints in doing research under the conditions of violent conflict is the sparse availability of micro-level data. No secondary data set about the socio-economic and livelihood activities at micro-level are available, and as Korf (2004) and Bohle & Fünfgeld (2007) stated, there are practical difficulties in conducting a large quantitative survey in a politically unstable condition due political and logistical constraints. It was only through methodological flexibility that was I able to gather the most relevant information in a very insecure environment. I used unstructured dialogues, conversations, interviews, and observations to collect the primary micro-level data required to capture the complex human behaviour and new insights into the processes involved in determining livelihood strategies of the different communities. I held extensive discussions with the Government Agent of the District, Divisional Secretaries, the district planning officers, the directors of various government departments, Sri Lankan army officers, home guards, village officers, religious leaders, leaders of community-based organizations (CBOs), academics of universities, representatives of traders, and, most importantly, with the war-affected rural households of all three communities. As many preferred to speak off the record, I have refrained from naming the persons concerned. Their knowledge, experiences, opinions, beliefs, needs, and constraint were all discussed and shared openly. In addition, documentary research was conducted to gather information on the macro-profile of the district and the context of the ethnic war. A summary of the methodological description of the study is given in Table 1.

The field approach started with purposive sampling, which was carried out at two levels. The first level of sampling started with a range of key stakeholders discussions in which all the villages affected by the war and/or conflict were identified and then three case study villages, a Sinhala village, a Tamil village, and a Muslim village, were selected. The selection was made to represent the different ethnic groups and the degree of conflict affectedness such a way as to shed light for the research objective of the study. The village names have been changed to protect the confidentiality and privacy of respondents and their respective villages. The background information relating to the three case study villages is summarized in Table 2.

At the second sampling level, a total of 21 households and three focus groups (seven households and one focus group from each of the sample villages) were selected according to their ethnicity, age, gender, and degrees of vulnerability. All of the in-depth interviews and informal discussions were

*Table 1.* Methodology

| Method | Size | | Composition | | Major Focus |
|---|---|---|---|---|---|
| In-depth interview | Total sample: | 21 HH | Main occupation | Number of HHs | Vulnerability and Livelihood strategies |
| | Sinhalese: | 7 HH | | | |
| | Tamils: | 7 HH | Agriculture | 4 S 2 T 2 M | |
| | Muslims: | 7 HH | Wage labour | 1 S 3 T 2 M | |
| | | | Monthly salary earners | 2 S – 1 M | |
| | | | Traders | – – 1 M | |
| | | | Others | – 2 T 1 M | |
| Focus group discussion | Total groups: | 3 | Farmers, Wage labourers, Traders, Fishermen, Self-employed people, Students, Widows, Teachers, Home guards | | Assets and Strategies |
| | Gr 1 SV | 6 Ma & 5 F | | | |
| | Gr 2 TV | 5 Ma & 7 F | | | |
| | Gr 3 MV | 5 Ma & 4 F | | | |
| Document research | – | | – | | Macro-profile, Methods, Framework, Relevant studies |
| Key informants discussion | Number of informants at a time | 1–5 | District Secretary, Divisional Secretaries, Village-level Administrators, CBO leaders, NGOs, Trade union leaders, District politicians, Directors of various public institutions, Teachers, Transporters, Defence officers, Ministries and Government institutions | | Context, Assets, Institutions, Policies, Identifying proper sample |

Key: HHs – Households, Gr – Group, SV – Sinhalese village, TV – Tamil village, MV – Muslim village, Ma – Male, F – Female, S – Sinhalese, T – Tamil, M – Muslim carried out by me during my research visits to the district. Two research assistants, one male and one female, from each of the communities in the local area helped in facilitating the group discussions. I took less of a directive role throughout the discussions and allowed the group dynamics to unfold with their comments, explanations and share experiences.

## The case study villages: the context and people's perceptions

### *Kotagala: Sinhala village*

'Kotagala' is a Sinhala settlement village located in the Uhana DS Division, bordered by Tamil and Sinhala villages, and has been subject to frequent attacks from the Tamil rebels in the past. This village is well connected to the nearest town by a metalled road of reasonable quality, and also by public and private transport. It has irrigated lands under the Gal-Oya scheme. Approximately 95% houses have electricity. The level of human capital, in general, is high. The general level of education is high for both sexes: more than 60% of the population has secondary education. The village children have access to education within the village up to university entrance level. The main form of social capital available in the village is the CBOs, comprising the Funeral Aid Society, Temple Society, Rural Development Society, Women's Development Society, and Farmers Society. However, the widely held view of the communities is that most of the CBOs are not considered to be of any significance to the community, due to frequent cases of corruption, favouritism and patronage of politically influential and wealthy members of the respective organizations.

The employment structure in the village is complex. In a sense, everybody is a small farmer, due to the fact that even those engaged in other types of employment (permanent employment in public and private sectors, trade, etc.) cultivate their own paddy lands. According to the village level administrator (Grama Niladhari or GN), almost 65% of the local households owned irrigable lands (approximately 12 families) in the localities, which consisted of large landholdings, and they owned vehicles such as tractors, vans and motorbikes. Paddy income is supplemented with income from other sources, including livestock keeping, the cultivation of other food crops, involvement in cottage industries, and employment in the public sector including the army and home guard service. Since the army and home guard provides attractive new sources of income to younger generations, there is intense competition among the youths in the village to sign up for such jobs. Women perform a substantial quantity of work at home and in the fields, yet the wages paid to women are less than those paid to men for carrying out similar work. Many of the women maintain a home garden. Some prepare cooked foods, especially rice flour based foods for sale. In addition to the aforementioned forms of employment, approximately 5% of females in the village engage in public sector work as teachers, clerical staff, etc.

Prior to the conflict, the livelihoods of the villagers also depended on slash-and-burn (*chena*) cultivation in the jungle area situated *c.*1 km from the village, and also the collection of herbs from the jungle to sell. Further, there was found to be well-developed ethnic mixing with other communities. This helped the local landless and poor casual workers to migrate to Batticaloa District, which is predominantly Tamil, and to Muslim villages in search of work:

> There was a peaceful environment till 1990. It started to change after that, and it turned worse in 1997. In the past we had a good relationship with the Tamil people who now live on the Tamil border. We used to travel to their areas and they came to our areas and we kept on interacting with them for a long time. (Respondent, 56 years, male)

The village underwent a number of attacks by the LTTE in the 1990s. Due to the repeated attacks, some farmers were displaced temporarily to other areas. Many people have sold their land and migrated to the non-conflict zones of the country due to the unstable security situation. The practice of slash-and-burn cultivation has been abandoned by the villagers due to continuous attacks by the Tamil rebels. The paddy lands close to jungle have also been abandoned due to the fear of possible attack by the LTTE. The loss of *chena* cultivation and ethnic segregation has had a direct negative impact on livelihoods. Today, most of the villagers have begun to cultivate their paddy land with the protection of the army and home guards. As in many other Sinhala villages in the conflict areas, the government offered the protection of armed forces:

> We are farmers. Our livelihoods were severely affected by the conflict. We were not allowed to cultivate the extreme corner of our land (bordering the Tamil village). During the peak of the conflict most of the villagers were in hiding the whole time and

*Table 2.* Background information relating to the three case-study villages

| Name & DS Division | Ethnicity | Experience of violence | Major income source(s) |
|---|---|---|---|
| Kotagala* Uhana DS | Sinhalese | The village was underwent attacks by Tamil militants a number of times in the 1990s. In one incident, more than 50 civilians were killed in during night-time. Sporadic violence continues. | Agriculture, Casual wage employment, Private/public sector employment, Home guard service |
| Sorakulam* Navithanveli DS | Tamil | During the SLA's round-up of the LTTE in 1990, many people were killed, disappeared or were displaced. Their homesteads were destroyed. Sporadic violence and harassment continues. | Agriculture, Casual wage employment, Remittances |
| Senkulam* Sammanthurai DS | Muslim | Experienced displacement due to attacks by Tamil militants in 1990. Experienced numerous incidences of killing, abduction, and harassment during IPKF occupation. Sporadic violence continues. | Agriculture, Casual wage labour, Private/ public sector employment, Commerce, Remittances |

* Village names have been changed to protect confidentiality

> the lands were just abandoned. In the beginning we worked during the day, around 4pm, and almost all of the villagers went to a common place at night. Those days it was rare to see a man walking after four in the evening. It was only us who did not even go and stay in a common place or at a friend's house for the night in this whole area. Now we are working without fear, I am sure the normalcy will return soon. We are sure that the army, police and home guards will take care of us. (Farmer, 43 years, male)

According to the paddy cultivators interviewed, the resumption of the conflict in December 2005 has had hardly any impact on paddy output in recent years. Since the conflict had destroyed their traditional supplementary income earning opportunities, some individual members of farming households had moved away from agricultural activities by diversifying their income sources into non-agricultural activities, yet still retained a strong connection to their traditional livelihood system:

> I worked in the paddy field with my father before joining the home guards. Even now, during my off-duty hours I work in the paddy field. But, it is difficult to earn enough through agriculture alone. This source of income does not allow us to manage day-to-day expenses, and to develop our living style to achieve better status. (Home guard, 24 years, male)

Accessibility of markets is a severe problem faced by farmers and traders in the village due to numerous security checkpoints along the major and minor roads throughout the district. All the respondent paddy farmers in the village opined that there was no easy access to places to market their produce and buy their inputs. In addition to restrictions on access, the villagers also identified the increasing cost of living as a major constraint to their livelihoods. In particular, poorer families found it much harder to cope with increasing costs of living. Such households explained that instead they have had to change consumption patterns, often by missing a meal or replacing the most common ingredients of a meal with less costly substitutes.

According to a village-level key informant, there is a high degree of unity amongst the households in the village, particularly during community events such as funerals and weddings. On such occasions, almost all voluntarily help those in greater need. However, in general, the village is highly politicized. The respondents reported political discrimination, as stated by one respondent: 'we have had to face political discrimination due to our party affiliations'. According to a village-level key informant, all parts of the village life are highly impregnated by party politics and power play.

The conflict has created livelihoods opportunities in home guard services for a section of Sinhala youths. Home guards are recruited from rural communities and given basic training, in the role to supplement the police and army, particularly in providing security to villages, access roads, and other civilian and military infrastructure. They are usually assigned work in the areas in which they live. Since 1993 home guards have been issued with automatic rifles. In April 2006 the Home Guard Service was renamed the Civil Defence Force. The number of those involved in home guard service had increased greatly in the past. In terms of employment, this option represents a stable permanent income in excess of LKR15,000 per month (USD1 = LKR108). This new income opportunity, coupled with other income sources, has enabled some households to improve the conditions of their houses and develop their income sources. Hence, the home guard service is considered by some villagers to be a favourable outcome of the conflict:

> Lots of young villagers have got good local jobs [as home guards] and no longer worry their parents by going off to Colombo and other places looking for work. They are well paid; they have job security and social status as never before. My two girls [daughters] are seeking employment [and want] to join the home guard. They are very happy with this alternative as the only other jobs are in the Free Trade Zones (Colombo). As home guards they get a good salary and a high level of social dignity, which was unimaginable before. A proper government job like this confers on them the highest possible status in the village. (Villager, 56 years, male)

In comparing living standards of those who cultivate land to those who work as home guards, it is apparent that the latter lead a relatively better life. The home guards also own televisions, cassette recorders and items of furniture that have been have been paid for in instalments.

### *Sorakulam: Tamil village*

'Sorakulam' is a traditional Tamil village located at the far end of the Gal-Oya irrigation scheme in Navithanveli DS Division (Ampara District), and is bordered by Muslim and Tamil villages. All households in the village were displaced during the conflicts in 1990 and sought refuge in the Tamil-dominated areas of the district and in Batticaloa. Almost every family had lost their property, social status and their traditional forms of livelihood due to conflict. The destruction of water channels during the conflict by villagers living upstream and the lack of maintenance forced the farmers to be totally dependent on the north-east monsoon rain for cultivation. To obtain drinking water, women have to walk *c.*1 km and sometimes men cycle to collect it. The road that leads to the village from the main road is a small gravel track. The closest bus stop is *c.*3 km distant from the village. The general level of education is low for both sexes, and no external efforts have been made to develop human capital in the village. However, the village youths are keen to improve their skills and capabilities in order to be competitive in the job market. The village school has classes up to secondary level (Ordinary Level). To study beyond secondary level, the pupils have to travel to Sammanthurai (5 km distant) or Kalmunai (7 km distant), but there are no transport facilities available for schoolchildren.

In a group discussion it was reported that during the period of displacement (when most people were relocated to welfare centres) local and international NGOs provided women, especially widows and families who had lost income earners, with the opportunity for income generation through micro- and small-scale enterprises. However, the lack of financial support to start micro-level projects and also the uncertainty regarding the security situation in the field compelled many women to depend on government remittances. Within a traditional conservative Tamil culture, women's decision making was generally confined to household matters, the education of children and health. However, the conflict has changed these limitations due to the loss of family members, death, and the disappearance of the main (mostly male) income earners.

Today, the main sources of livelihood in the village are paddy cultivation, remittances and supply agricultural labour to paddy producing areas. Men leave their homes early in the morning for casual work and return in the evening. Although the role of women had expanded due to the conflict, most of their occupations could be termed survival occupations (with the exception of foreign migration), and do not contribute much towards restoring assets. The marketing of local agricultural produce is usually done by women. Some engaged in activities such as sewing, weaving and petty trade. The women and children of poorer families sometimes take high risks by collecting firewood in the jungle. Most of the women in the village are involved in fishing, mainly for consumption but any surplus is salted and dried. According to a key informant, brewing illicit liquor is also practised by some villagers as a livelihood strategy.

During the group discussion, the villagers reported that their lives have marginally improved since the MOU, but they continue to feel a deep sense of discrimination by the government regarding the degree of access to economic opportunities. In 1990, they were driven out of their villages. When the SLA moved in, some families took refuge in the nearby jungle and others in camps due to fear of arbitrary arrests, looting and rape. Recently, they have been taken to welfare centres at Thirukkovil, a village located in the east coast of Sri Lanka that was hardly affected by the tsunami in December 2004. A large number of villagers have been either killed or have disappeared in the worst of riots and inter-ethnic conflict. Months of attacks, and associated looting and destruction of homesteads, have led to a massive loss of core livelihood assets, including the loss of income earners, livestock and permanent crops. A fraction of the displaced Tamil people managed to return to their villages in 1991, but most of the displaced people returned only after the signing of the MOU in 2002, due to their inability to create stable livelihoods in other areas. It was revealed in the focus group discussions that a large number of youths had not returned following displacement, but were staying in Thirukkovil and Pottuvil (both on the east coast), where they were involved in various economic activities:

> In 1990, my family fled this area with only with our clothes, leaving everything behind. We lost everything. Our house was fully destroyed. We didn't get enough compensation. We sold everything that we had, including my wife's wedding chain (*Thali*). When we returned, we had nothing to sell to survive except our paddy land. (Male, 49 years)

During the period of war, an interesting feature of the landownership pattern was the transfer of ownership of some of the land from Tamils to Muslims. Due to economic hardships and uncertainty concerning the future, large numbers of Tamil farmers sold their lands to Muslims. The villagers' livelihoods had been severely affected as a result of the conflict, and they were not in a frame of mind to think of earning money or making profit; it only mattered that they saved their lives in the short term by selling land.

In the group discussion and interview in the village it was revealed that prior to the MOU of 2002, the Tamils, especially the youths, feared leaving their village due to possible harassment by the SLA and the Tamil rebels. During the period 2002–2004, all the villagers were able to move freely and without fear, but today the situation has changed once again. As with the situation prior to 2002, they have clearly recognized freedom of movement to be a critical part of their ability to live and work. Many of those interviewed reported that one of the major problems that they faced was the checkpoints of the SLA and Tamil militant groups. Passing through checkpoints for their livelihood activities is highly risky, particularly for young males and females, and even in the current situation of no open warfare Tamil farmers do not want to risk passing through checkpoints and Sinhala villages. According to one key informant, whenever the LTTE attacked the SLA, the troops attacked the adjoining Tamil villages in revenge. Such attacks are becoming more individualized and targeted at households for personal gain rather than being politically motivated:

> To go out of the village is a problem. Young men and women here have no freedom due to the conflict situation. They always are facing problems when they go out. We do not have confidence to go to the field alone or to go two to three miles away from the village. (Female, 32 years)

Forest areas have become completely insecure places for the villagers. In the interview, it was revealed that before the war, collecting and selling firewood and other raw materials constituted some of the supporting livelihood activities practised by female members of poorer households. However, conflict has rendered such livelihood activities a major threat:

> Not like during the pre-conflict period, we now have less option to make our livelihood. Large numbers of our men were killed throughout the war. The biggest problem for their widows is trying to support themselves and their children. Even if they do manage to find some source of income generation, people in our community often assume that this money must come from prostitution. (Representative of a women's society, 47 years, female)

The conflict has led to high levels of migration away from the affected areas, especially among the Muslim and Tamil communities. In many Tamil families migration was considered more as a survival strategy than an income generating strategy. It was revealed in the group discussion and the discussion with the village-level key informant that the conflict paved the way for some people to move to foreign countries, especially to Middle East and Western countries:

> Men and young girls, more than women, who fear recruitment by the militants or being arrested by SLA, have been forced to migrate. Due to conflict only old men, women and children are mostly left in villages. (Villager, 46 years, female)

One of the features of conflict-related migration in Sri Lanka is that people in conflict areas have been able to access Western countries by means of temporary migration because of special considerations in place for conflict-affected people.

The village is lacking political and social networks. It was revealed in the group discussion that resource allocations, such as land distribution, distribution of housing and financial supports, etc., are generally under the control of national government ministers and the ruling party members of parliament, and therefore are distributed with a strategic and political agenda. Thus, some people use their political networks not only to secure resources but also to secure government employment:

> There are so many educated youths without employment. Our youths do not know Sinhala. So it is difficult for them to get jobs elsewhere. Even if they want to go and work outside the district, the situation is unsafe for them. Education benefits only the rich to find jobs. The poor, having spent years for a degree, have to wait at home without a job. This happens because most of the jobs are under manipulation from the government. Even people without educational qualifications are sitting in high positions and spoiling the institutions. (Graduate, 28 years, male)

The Tamil community suffers from a lack of political representation in the ruling alliance. Since government employment was one of the few permanent income sources that continued during the conflict period, the lack of opportunity to secure government employment due to lack of political networking strongly affected the Tamil community in conducting their lives in the conflict situation. It was also revealed that the social networks outside the village are restricted by security concerns, which inhibit freedom of movement, particularly for girls and young women who often are not allowed to move around freely. Such restrictions were heavily enforced upon young women in the village. Also, lack of transport was a major impediment to maintaining relationships with people outside the village.

With regards to water resources in the village (as well as in the district), much of the agriculture depends on the Gal-Oya water project. The scheme provides water upstream to (Sinhalese) settlers, yet Tamil and Muslim farmers from the area had received land at the tail end of the scheme. Such policy provides the upstream farmers with a strategic advantage in control over water resources and water flows in the irrigation schemes. It was revealed during the field study that the upstream farmers divert more water for their cultivation than was allocated to them, thereby creating overall water shortages or delayed water delivery for downstream farmers. It was further reported that when upstream farmers blocked the flow of water to divert more water to their fields, the lower stream farmers had very little scope to intervene in any way due to the intimate links between the upstream farmers and the SLA.

## *Senkulam: Muslim village*

'Senkulam' is a densely populated Muslim resettled village located in Sammanthurai DS Division. The village is bordered by Tamil and Muslim villages, and experienced a brief period of displacement due to the attacks by Tamil rebels in 1990. During the occupation of the IPKF, the village experienced numerous incidents of killing, abduction, and harassment. However, the people enjoy comparatively better transport, a school, and health care facilities due to the village's proximity to Sammanthurai town. The village possesses high human and social capital. The houses in the village are of good quality and have electricity supplies. There is easy access to education up to university entrance level and also good health care facilities. The employment structure in the village is quite diverse as no single work category is predominant. People are engaged in number of micro-enterprises, trade, farming, fishing, and employment in the public sector. Some women are engaged in public sector work, mainly as teachers. Also, some women occasionally earn cash but are dependent primarily on male earnings, while very poor women engage in home-based informal sector activities such as weaving, rice pounding, poultry keeping, and dress making. It was revealed that the direct long-term impact of the conflict on the employment structure in the village has been insignificant: 'Even before the conflict started we were engaged in similar livelihood activities as we are engaged in now. All activities were affected by the conflict in 1990. However, after the conflict we started the same again' (male respondent, aged 43 years).

Prior to the present ethnic problem, there were relaxed and reciprocated social relationships with other communities, particularly with Tamils. The villagers used to move freely in Tamil and Sinhala areas, and similarly the Tamils and Sinhalese visited this village. This helped them to readily access various markets, including the labour market, but during the conflict relationships deteriorated, especially with Tamil villages. Distrust has resulted in a process of ethnic conflict between these two communities, and as a direct consequence both communities have become highly segregated. Today, there little interaction between the groups and mutual fears of being harassed are widespread.

The large numbers of Muslims living in the coastal belt of the district have continued to face shortages of land necessary for agricultural activities. There is no state or private land available in these areas. Although the village is in the vicinity of a market town (like most Muslims villages) and is considered a safe place, the paddy land is located outside these safe places and adjacent to fields belonging to Tamils, hence Muslim farmers have to cross Tamil settlements and fields in order to work in their own fields. This creates a regular process of tense interaction between the two communities. Many people depend on such interactions for their livelihoods; they have bought paddy land owned by Tamils in the past. On the other hand, they have also lost many areas of paddy farming that were scattered among Tamil areas due to restrictions imposed by the LTTE, as stated in the group discussion:

> We (Muslims) have bought some of the Tamils' lands. They sold the land of their own will. But immediately after the MOU they came with persons from their militant groups and asked for the land back. From 1987 we had many threats from Tamils militant groups. The IPKF, and EPRLF [Eelam People's Revolutionary Liberation Front] were responsible for causing trouble earlier. After 1990, the LTTE infiltrated into the southern part of Ampara (particularly in Thirukkovil) and brought a large portion of land belonging to Muslims under its control. We [the displaced Muslims] underwent number of hardship during the last 20 years.

> Now after the GoSL cleared these areas of the LTTE, the TVMP [Tamil Makkal Viduthalai Pulikal or Tamil Peoples Liberation Tigers] with the help of the government security forces is putting up permanent houses for Tamils on our land without any consideration for Muslims' right to settle the displaced Muslims. (Displaced farmer, 53 years, male)

Few villagers expressed their anger about the land that once was theirs but had since been claimed by the Sinhalese. They claimed that the situation had been compounded by local government officials and politicians who were biased against the Muslims. Sinhala Buddhist nationalists claim the Ampara District traditional lands as their heritage on the basis that there are many ruins of Buddhist shrines, yet the Sinhalese acquire the lands claiming that they rightfully belong to their temple (*Vihara*). One respondent stated:

> Even if we are given land rights they might say that it belongs to their temple and ask us to leave. . . . We have had to face political discrimination due to our political (party) affiliation. (Farmer, 34 years, male)

There are number of internally displaced people in the village, and they believed that the LTTE has been the core cause of their displacement. All reported the positive functions of the social network within their village. For them, displacement caused deep changes in their livelihood practices. They still relied primarily on the support mechanisms within the Muslim community, including kinship networks, Islamic organizations, and charity from the mosques. However, some reported the lack of functioning kinship networks due to the geographical distance between siblings:

> We were chased from our land [in Batticaloa] by the Tamil militants in 1990. The income earned from casual work is hardly enough to survive. We are given some bags of paddy every year by the mosque. We normally sell a portion of that and buy other necessities. We lost all our assets in 1990 and after that we couldn't get ahead. That was a major break-up in our life. We are unhappy about our current situation, and still suffer that loss. (Respondent, 42 years, female)

A few villagers have a powerful role in trading. Their good relationship with the SLA and police enables them to transport goods through checkpoints. This gives them powerful leverage in their relations with Tamils. However, one trader respondent indicated that accessibility to market remains a severe problem faced by traders due to numerous security checkpoints along the major and minor highways throughout the district. Those manning the checkpoints are often corrupt, and ransom and bribes are taken by some parties. The restrictions tend to increase prices, and hence the costs of maintaining a livelihood.

Lack of security in obtaining a reliable income and diversification of livelihood activities were identified as two of the main obstacles to maintaining what the villagers considered to be a reasonable standard of living. The ability to develop the skills necessary for work, to migrate, and to find a remunerative job, were also seen as important: 'We have to get good education to go to foreign countries' (male youth respondent, 23 years).

Conflict has created a high degree of out-migration. Labour migration, especially to Arab countries is a predominant source of livelihood for fairly a large number of people in the village, and this is considered more as an asset building strategy than a survival strategy. As Brun (2003) stated, international labour migration is considered one of few possibilities for the people to accumulate sufficient money to be able to buy a plot of land and build a house, and hence restore their living standards. Many families in the Muslim community have become better off in economic terms after at least one member of the family started migrating to the Middle East. In a focus group discussion with Muslim people in the village, it was claimed that labour migration and remittances were the most important source of income for a large number of families:

> Before the conflict we had enough land in Mannar. But we have lost our properties due to the conflict, and came to this settlement without any assets. I manage to run my family by working as a wage labourer. What crime did we commit to deserve this punishment? Now we don't have any land to cultivate. I have five children [two boys and three girls]. My eldest son, after finishing his Ordinary Level exam, has gone to Dubai with the help of a relative who is working there. Our main source of livelihood at present is his income from abroad. His income helps us to have some savings which is sorely needed for the purpose of my daughters' marriages. (Resettled farmer, 59 years)

Labour migration allows people to diversify their income sources and thereby increase their resilience in periods of hardship. Hence, migration has to be considered as a most significant factor in the wider context of people's livelihoods, in their strategies to diversify sources of income and in terms of reducing risk.

## Analysis of case studies

Conflicts destroy livelihoods, either directly through causing death and destroying property, or indirectly by limiting freedom of choice and adaptive behaviour, thereby hampering the pursuit of successful survival strategies. Although the government declared that the Eastern province has been liberated from the LTTE, the people in the Ampara District clearly regard freedom of movement as critical aspect of their ability to live and work. Many people reported that they were unable to access markets, or farm land in other areas, or to fish where they wanted because they feared for their safety. Even today, freedom of movement is far from complete, with many reporting they do not feel safe in all locations. Many of those working in trades, agriculture and fishing reported that significant threats were still made to their personal security when they attempted to travel and work outside their home area. It has been observed that for most people fear had a significant effect on their behaviour. The villagers generally felt that the government should provide more for their security, but they had virtually no confidence that the government could deliver in this respect. This sentiment is linked to the unstable political situation in the district. Many said they felt constrained in expressing their opinions to people outside their own family or to supposedly reliable persons. Some, especially Tamils, were concerned that freely expressing their opinions could lead to retaliation if a riot or other conflict were to occur. However, many of those interviewed reported that there is less tension between the ethnic groups than in the past, but it is clear that there are still are deep rifts between communities. For many, the events of the past two decades have seriously eroded their ability and willingness to trust members of other ethnic groups.

The prolonged conflict restricts access to human capital in a number of ways. In rural areas there has been a loss of public infrastructure, including the destruction of schools, health centres, and transport services. The availability of public services has been further decimated, as civil servants, teachers, health workers, and others have either become displaced or have decided not to remain in the affected areas. The breakdown of health and education services provided by the state and private sectors has restricted the development of human capital. Qualified doctors and teachers are reluctant to work in the district mainly due to poor security. Most of the villages in the interior are not connected with townships by reasonably maintained roads and by public transport. In most cases, people either walk or use bicycles. Facilities available for medical services are non-existent for a large number of villages. In all three communities, most of the parents do not have enough money to spend on their children's education.

The study shows that the villagers adopted various livelihood strategies in order to cope with the conflict situation. Some of the livelihood strategies observed in the three study villages are summarized in Table 3. The families in all the three locations, including those who suffered their day-to-day survival even before the conflict, found it much harder to cope with the situation which blocked their limited income options through the restrictions of the labour market. The conflict created a class of the seriously poverty stricken, due to the destruction of their homesteads and or loss of primary income earners. They coped with the situation mainly by selling their remaining tangible assets in order to survive and by reducing the number, size and quality of meals.

In the Tamil village, households are very reluctant to build up tangible assets such as investments in land or making repairs and improvements to their houses due to uncertainty about the future. However, in contrast, in the Sinhala and Muslim villages it was observed that people were taking some level of risk by investing in land and building new houses in attempts to rebuild their future.

As a survival strategy during the eruption of violence, conflict-affected villagers adapted their geographical action spaces. Retreating to the jungle and moving from there to Tamil-dominated geographical spaces was an immediate survival strategy adopted by the Tamils. In contrast, the Sinhalese and Muslims developed social networks as an immediate survival strategy, either within their area of domicile or by moving to the non-conflict zones of the country (Sinhalese village) or to safer places within the conflict zone (Muslim village). There was a tendency during the peak of the conflict to reduce physical assets in the conflict areas and to build them up elsewhere in non-conflict areas.

Foreign migration is used as a strategy in the Tamil and Muslim villages to become economically better off, and also as a survival strategy by the Tamils (by living in safer places). In the Sinhala village, many young men and women work as home guards in order to become economically and socially more affluent. The livelihoods of the home guards are mainly dependent upon the conflict environment, but in a post-conflict situation there will be uncertainty about the sustainability of this form of livelihood for youths. Further, the attraction of the job discourages many youths from continuing with higher education, who leave school before completing secondary education. Hence, in future, this strategy may reduce the human capital of the community.

In addition to the aforementioned strategies, it was also observed that new linkages with NGOs, more flexible gender roles, and borrowing small group loans were some of the livelihood strategies used in the Tamil village. Some of the common strategies used in the Muslim village included

*Table 3.* Major livelihood strategies practised in the three case study locations

| Major livelihood strategies in the Sinhalese village | Major livelihood strategies in the Tamil village | Major livelihood strategies in the Muslim village |
|---|---|---|
| • Developing social networks within the village during the eruption of violence<br>• Changing geographical space temporarily to safer areas<br>• Avoiding crossing Tamil villages<br>• Joining home guards<br>• Working in groups and under the provision of security by SLA and home guards<br>• Coping with local level power holders<br>• Having flexible gender roles<br>• Paddy cultivation and cultivation of subsidiary food crops<br>• Reduction in number of meals<br>• Investing in tangible assets such as land, house and household items<br>• Forming CBOs to gain access of welfare and development | • Fleeing to the jungle and/or Tamil-dominated areas during eruption of fighting<br>• Residing in the welfare camps or with relatives in safer areas<br>• Avoiding sending men and young women through check-points, crossing villages of other communities, or going out after dusk<br>• Females managing household economics; handling flexible gender role.<br>• Involving agriculture, working as casual wage earners, and involving in micro-enterprises<br>• Sale of assets, taking out loans; coping with NGOs<br>• Sending children away; foreign migration<br>• Working in the field and marketing in groups<br>• Reducing the amount, type and number of meals | • Changing geographical space temporarily in the conflict zone<br>• Avoiding crossing Tamil villages and going out after dusk<br>• Developing social networks within the village to face the problems of conflict<br>• Migration to Middle East countries<br>• Reducing number of meals<br>• Investing in children's education<br>• Paddy cultivation, conducting business, petty trade, and working as casual wage earners<br>• Investing in land, housing, household items, and jewellery<br>• Establishing good relationships with power holders |

developing social networks among Muslims, borrowing from relatives and friends, and coping with domestic-level power holders. In contrast, in the Sinhala village some of the common strategies included taking out loans, establishing good relationships with power holders, and having flexible gender roles.

## Conclusions

Conflict has affected the livelihoods of the people in the three study locations of Ampara District in different ways. Since the mid-1980s, the Tamil community has faced harassment from the Sri Lankan armed forces and also sporadic threats from Muslims, the Sinhala community has faced sporadic attacks from the LTTE, and the Muslim community has faced frequent attacks from Tamil militants. Further, the conflict has had a number of impacts on people's livelihoods through the restrictions on access to various types of assets and markets; the looting of properties, destruction of infrastructure, indiscriminate attacks on civilian areas, and general breakdown of law and order.

The basis of rural livelihoods in Ampara District is relatively simple to understand, as all ethnic groups are dependent on paddy cultivation to varying degrees. Most groups supplement their income from paddy cultivation with government remittances, and through rearing livestock, home gardening, fishing, trade, joining military forces, involving in micro-ventures, and foreign migration. Vast numbers of Sinhala families are heavily dependent on the armed forces to maintain or even raise the standard of their livelihoods. The standard of living of many Tamil and Muslim families is largely determined by migration to other countries. The comparative assessment indicates that most of the livelihood strategies adopted by people in all three locations studied are mainly context based. The findings also indicated that although most of the households are the losers in the conflict, there are a few winners, even at the micro-level.

The study revealed that the most fundamental problem affecting livelihoods in the district is insecurity, which restricts the mobility of all groups to varying degrees. Insecurity is widespread as a result of continued violent conflict between the SLA and Tamil militants. Limited mobility resulting from insecurity has seriously limited the core livelihood strategies of all communities to varying degrees. The villagers also face problems relating to land (especially Muslims), water (especially Tamils) and lack of welfare, and also ethnical segregation (all groups). Further, there is increased vulnerability of women, both economically and socially. It has been observed that for most people fear had a highly significant effect on their behaviour. Finally, single women whose husbands have fled or been killed, bear a double burden of fear.

*Acknowledgements.* – The author is grateful for useful and detailed comments from Professors Piers Blaikie, Ragnhild Lund, Cathrine Brun, and two anonymous referees.

## References

Abeyratne, S. 2004. Economic roots of political conflict: The case of Sri Lanka. *The World Economy* 27:8, 1295–1314.

Bastian, S. 1985. *University Admission and the National Question in Ethnicity and Social Change.* Social Scientists' Association, Colombo.

Bhatia, M., Goodhand, J., Atmar, H., Pain, A. & Suleman, M. 2003. Profits and poverty: Aid, livelihoods and conflict in Afghanistan. Collinson, S. (ed.) *Power, Livelihoods and Conflict: Case Studies, in Political Economy Analysis for Humanitarian Action.* HPG Report 13. http://www.odi.org.uk/hpg/papers/hpgreport13.pdf (accessed September 2008).

Blaikie, P.T., Cannon, I., Davis & Wisner, B. 1994. *At Risk: Natural Hazards, People's Vulnerability and Disasters.* Routledge, London.

Bohle, H.G. & Funfgeld, H. 2007. The political ecology of violence in eastern Sri Lanka. *Development and Change* 38:4, 665–687.

Brun, C. 2003a. *Finding a Place: Local Integration and Protected Displacement in Sri Lanka.* Dr.Polit Thesis, NTNU, Trondheim.

Brun, C. 2003b. Permanent impermanence: Dilemas of long-term displacement in Sri Lanka. Brun, C. & Birkeland, N.M. (eds.) *Researching Internal Displacement: State of the Art.* Acta Geographica (Trondheim), Series A, No. 6, 423–432.

Brun, C. 2008. Birds of Freedom: Young people, the LTTE, and representation of gender, nationalism and governance in northern Sri Lanka. *Critical Asian Studies* 40:3, 399–422.

Carney, D. 1998. *Sustainable Rural Livelihoods: What Contribution Can We Make?* Department for International Development, London.

Chambers, R. & Conway, G. 1992. *Sustainable Rural Livelihoods: Practical Concepts for the 21st Century.* IDS Discussion Paper 296. Institute of Development Studies, University of Sussex, Brighton.

Chandrakanthan, A.J.V. 2000. Eelam Tamil nationalism: An inside view. Wilson, A.J. (ed.) *Sri Lankan Tamil Nationalism: Its Origins and Development in the Nineteenth and Twentieth Centuries,* 157–173. University of British Columbia Press, Vancouver.

Collinson, S. (ed.) 2003. *Power, Livelihoods and Conflict: Case Studies in Political Economy Analysis for Humanitarian Action.* HPG (Humanitarian Policy Group) Report 13, 7–19. Overseas Development Institute, London http://www.odi.org.uk/hpg/papers/hpgreport13.pdf (accessed January 2008).

DCS. 2007. *Basic Population Information on Ampara District.* Department of Census and Statistics, Colombo.

de Silva, C.R. 1984. Sinhala-Tamil relations and education in Sri Lanka: The first phase, 1971-77. Goldmann, R.B. & Wilson, J. (eds.) *Independence to Statehood: Managing Ethnic Conflict in Five African & Asian States,* 125–146. St. Martin's Press, New York.

De Waal, A. 2005. *Famine that Kills: Darfur, Sudan 1984-85.* Oxford University Press, Oxford.

Ellis, F. 2000. *Rural Livelihoods and Diversity in Development Countries.* Oxford University Press, Oxford.

Fernando, L. 1997. Ethnic conflict and the state in Sri Lanka: A possible solution? *South Asia: Journal of South Asian Studies* 20:1, 83–96.

Gamage, S. & Watson, I.B. 1999. *Conflict and Community in Contemporary Sri Lanka.* Sage, New Delhi.

Gilbert, R. 2001. *Doing More for Those Made Homeless by Natural Disasters.* The World Bank, Washington DC.

Goodhand, J., Hulme, D. & Lewer, N. 2000. Social capital and the political economy of violence: A case study of Sri Lanka. *Disasters* 24:4, 390–406.

Hasbullah, S., Balasundarampillai, P. & Silva, K.T. 2005. *Addressing Root Causes of the Conflict: Land Problem in North-East Sri Lanka.* UNHCR and FCE, Colombo.

Hellmann-Rajanayagam, D. 1990. The politics of the Tamil past. Spencer, J. (ed.) *Sri Lanka: History and the Roots of Conflict,* 107–124. Routledge, London.

Human Rights Watch. 2007. *Human Rights Watch, Return to War: Human Rights under Siege.* http://hrw.org/english/docs/2007/08/03/slanka16573.htm (accessed May 2008).

Jaspers, S. & Shoham, J. 2002. *A Critical Review of Approaches to Assessing and Monitoring Livelihoods in Situations of Chronic Conflict and Political Instability.* Working Paper 193. Overseas Development Institute, London.

Jayawardena, K. 2003. *Ethnic and Class Conflict in Sri Lanka: The Emergence of Sinhala-Buddhist Consciousness 1883-1983.* Sanjiva Books, Colombo.

Korf, B. 2004. War, livelihoods and vulnerability in Sri Lanka. *Development and Change* 35:2, 275–295.

Korf, B. & Fünfgeld, H. 2006. War and the commons: Assessing the changing politics of violence, access and entitlement in Sri Lanka. *Geoforum* 27:3, 391–403.

Korf, B. & Silva, K.T. 2003. *Poverty, Ethnicity and Conflict in Sri Lanka: Chronic Poverty and Development Policy.* University of Manchester, Manchester.

Korf, B., Flämig, T., Schenk, C., Ziebell, M., Ziegler, J., Singarayer, R.M., Abeyrathne, R.M., Devarajah, K., Dharmarajah, D. & Sakthivel, T. 2001. *Conflict-Threat or Opportunity? Land Use and Coping Strategies of War-affected Communities in Trincomalee, Sri Lanka.* Schriftenreihe des Seminars für Ländliche Entwicklung (SLE). Humbolt University, Berlin.

Lautze, S. 2007. *Best Practices and Lessons Learned: Rethinking Protection in Conflict Zones Using Livelihood Analysis.* UNICEF, Geneva. http://www.livesandlivelihoods.com/files/33594416.pdf (accessed September 2008).

Lautze, S. 2008. Social dynamics in militarised livelihood systems: Evidence from a study of Ugandan army soldiers. *Journal of Eastern African Studies* 2:3, 415–438.

Lautze, S. & Roberts, A.R. 2006. Violence and complex humanitarian emergencies: Implications for livelihoods models. *Disasters* 30:4, 383–401.

Muggah, R. 2008. *Relocation Failures: A Short History of Displacement and Resettlement in Sri Lanka.* Zed Books, London.

PTRR. 2005. *Post-tsunami Recovery and Reconstruction: Progress, Challenges and Way Forward.* Joint Report of the Government of Sri Lanka and Development Partners. PTRR (Post-tsunami Recovery and Reconstruction), Sri Lanka.

Reardon, T., Delgado, C. & Matlon, P. 1992. Determinants and effects of income diversification amongst farm households in Burkina Faso. *Journal of Development Studies* 28:2, 264–296.

Rennie, K.J. & Singh, N.C. 1996. *Participatory Research for Sustainable Livelihoods: A Guidebook for Field Projects.* International Institute for Sustainable Development, London.

Rogers, J.D., Spencer, J. & Uyangoda, J. 1998. Sri Lanka: Political violence and ethnic conflict. *American Psychologist* 53:7, 771–777.

Ruwanpura, K.N. 2006. Conflict and survival. *Asian Population Studies* 2:2, 187–200.

Schafer, J. 2002. *Supporting Livelihoods in Situations of Chronic Conflict and Political Instability.* Working Paper. Overseas Development Institute, London.

Seneviratne, H.L. 1999. *The Works of Kings: The New Buddhism in Sri Lanka.* University of Chicago Press, Chicago.

Shanmugaratnam, N. 2008. The dynamics of livelihood in a war zone: Mapping socio-economic change in eastern Sri Lanka. Shanmugaratnam, N. (ed.) *Between War and Peace in Sudan and Sri Lanka*, 117–148. James Currey, Oxford.

Shastri, A. 1990. The material basis of separatism: The Tamil Eelam movement in Sri Lanka. *Journal of Asian Studies* 49:1, 58–77.

Sidharthan, M. 2003. Negotiating Tamilness. A case study of Jaffna Tamil migrants to Colombo since 1990. Mayer, M., Rajasingham-Senanayake, D. & Thangarajah, Y. (eds.) *Building Local Capacities for Peace: Rethinking Conflict and Development in Sri Lanka*, 305–320. Macmillan, New Delhi.

Silva, K.T. 1999. Caste, ethnicity and problems of national identity in Sri Lanka. *Sociological Bulletin* 48:1–2, 201–215.

Silva, K.T. 2003. Armed conflict, displacement and poverty trends in Sri Lanka: Evidence from selected displaced populations. Mayer, M., Rajasingham-Senanayake, D. & Thangarajah, Y. (eds.) *Building Local Capacities for Peace: Rethinking Conflict and Development in Sri Lanka*, 245–270. Macmillan, New Delhi.

Snodgrass, D.R. 1999. The economic development of Sri Lanka: A tale of missed opportunities. Rotberg, R.I. (ed.) *Creating Peace in Sri Lanka: Civil War and Reconciliation*, 89–107. Brookings Institution Press, Washington DC.

Solesbury, W. 2003. *Sustainable Livelihoods: A Case Study of the Evolution of Department for International Development Policy.* Overseas Development Institute, London.

Stites, E., Lautze, S., Mazurana, D. & Anic, A. 2005. *Coping with War, Coping with Peace: Livelihood Adaptation in Bosnia, 1989–2004.* US Agency for International Development, Washington, DC.

UTHR. 1990. *The War and Its Consequences in the Ampara District.* Special Report No. 3. University Teachers for Human Rights (Jaffna), Jaffna.

Uyangoda, J. 2005. *Conflict, Conflict Resolution and Peace Building.* Department of Political Science and Public Policy, University of Colombo, Colombo.

Wilson, A.J. 2000. *Sri Lankan Tamil Nationalism.* Hurst & Company, London.

Wisner, B., Blaikie, P., Cannon, T. & Davis, I. 2004. *At Risk: Natural Hazards, People's Vulnerability and Disasters.* Routledge, London.

Young, H. & Osman, A. M. 2006. *Challenges to Peace and Recovery in Darfur: A Situation Analysis of the Ongoing Conflict and its Continuing Impact on Livelihoods.* www.scribd.com/doc/2094971/Annual-Report-2007 (accessed June 2008).

# 'Your subject of protection is a dangerous one': Protracted internal conflict and the challenges for humanitarian agencies

HAZEL LANG & ANITA KNUDSEN

Internal displacement in Sri Lanka exists in the context of a protracted condition of insecurity produced by intractable war and polarization of society along ethno-political lines. People remain exposed to risks and threats to their security over prolonged periods of time and the mobilization of inter- and intra-ethnic tensions exacerbates the breakdown of trust and protection at community level. Humanitarian agencies seeking to assist displaced civilians are compelled to grapple with challenges about how to engage in the most skilful and effective way in a risky and difficult environment. Drawing on field interviews in Trincomalee District, the article examines humanitarian agency approaches to protection in this politically challenging context. It is divided into six sections – international policy discussion on protracted internal displacement situations; the intractability of war in Sri Lanka; humanitarian approaches to protection; the political contours of the conflict in Trincomalee; the specific protection challenges confronting humanitarian agencies as they work in this highly politicized and militarized arena; and finally, the lessons and creative strategies crafted by agencies to negotiate these challenges.

*Hazel Lang, Key Centre for Ethics, Law, Justice and Governance, Griffith University*

## Introduction

Internally Displaced Persons (IDPs) remain embedded in the conflict that caused their displacement. Far from finding themselves in a place of safety, IDPs still carry most of the burden of a protracted conflict. As the UN Security Council's report on the Protection of Civilians in Armed Conflict (2007, 2) highlights: 'displacement ... continues to be one of the principal features of contemporary conflict and arguably the most significant humanitarian challenge that we face'. Humanitarian agencies seeking to assist and protect civilians displaced within a state and otherwise rendered insecure by prolonged, sometimes pulsating, civil war are confronted with challenges deeply connected with the political problems underpinnings the conflict. Agencies are compelled to grapple with challenges of how to engage in the most skilful and effective way in a risky and difficult environment. A politicized context is not the exception but the rule. Sri Lanka is regarded as among the most dangerous places in the world for aid workers (IRIN 2007).[1] The findings of field research detailed in this article are necessarily context specific but some of the issues and their implications have wider resonance with challenges and lessons for other difficult politicized protection crises and humanitarian operational environments elsewhere.[2] There is 'growing consensus within the international community, and the development aid community in particular, that disengagement from protracted crises, including conflict-affected states ... is no longer an option' (Macrae & Harmer 2004, 1).

Intractable war, political violence and polarization of society along ethno-political lines have generated a protracted condition of human insecurity in Sri Lanka. This chronic insecurity has produced continuous multiple, fluid and hidden forms of displacement and consequences, including the breakdown of protection and trust at community level. Drawing on field interviews in Trincomalee District, this article examines humanitarian agency approaches to protection in a politically challenging and insecure context. Following this introduction, the article is then divided into six sections – international policy discussion on protracted internal displacement situations; the perpetuality of war in Sri Lanka; humanitarian approaches to protection; the political contours of the conflict in Trincomalee; the specific protection challenges confronting humanitarian agencies as they work in this highly politicized and militarized arena; and finally, the lessons and creative strategies crafted by agencies to negotiate these challenges. First, we begin by outlining current international policy discussions on protracted internal displacement situations.

## Internal displacement in protracted crises

Particular attention to the issue of 'protracted IDPs situations' is a new area of policy discussion and has been the subject of relatively little analysis and action to date (Brookings-Bern 2007a). However, the majority of IDPs in the world today find themselves in a persistent state of limbo. Further, the problem is a large-scale one with two to three times more IDPs than refugees living in protracted situations (Brookings-Bern 2007a, 1, 39). Sri Lanka ranks highly as one of the major large-scale, long-term IDP-producing countries globally (Centre for Policy Alternatives 2007, 4).

As with refugees, protection for IDPs ultimately entails ensuring a durable solution to their plight (Brookings-Bern 2007b, 1). The question of when displacement ends, however, is not always clearly identifiable, as cessation is 'contingent upon a change in the factual situation of displacement' (Brookings-Bern 2007b, 17). The Guiding Principles on Internal Displacement (UNHCR 1998) deliberately do not contain a cessation clause for this reason, although Principles 28 and 29 do spell out conditions for voluntary return in safety and dignity to place of origin, resettlement and reintegration. 'Resettlement', as it is termed by the Guiding Principles, means local integration in the areas in which IDPs initially take refuge or relocation to another part of the country (IASC 2007a, 2). In developing a normative framework for durable solutions for IDPs, the Brookings-Bern Project on Internal Displacement defines a durable solution as 'no more displacement-specific needs and vulnerabilities' (2007b, 14), although 'this does not mean that they may not continue to have a need for protection and assistance, but their needs would be no different from other similarly situated citizens' (2007b, 8). Because conflict-induced IDPs remain citizens of the country in which they are living, solutions – at least theoretically – 'must be grounded in State responsibility whether or not the government is part of the cause of the displacement' (Brookings-Bern 2007a, 7). National responsibility for protection and solutions is particularly difficult when the state is unable or unwilling to protect its citizens and is itself a source of human insecurity. In its own interests, a government may claim an end to the existence of IDPs in an effort to create an image of normality to distract international scrutiny (IASC 2007a, 1) or, as in the case of Sri Lanka, to consolidate military or political objectives. Protracted internal displacement crises leave people exposed to risks and threats to their security and livelihoods over a prolonged period of time. The achievement of protection and solutions is always politically contingent – solutions may be elusive for IDPs due to threats, risks and vulnerabilities that continue to persist following putative or politically fragile efforts to resolve a crisis.

## Displacement and the 'incurable habit of war' in Sri Lanka

Uyangoda (2007, 4) explains that the parties to Sri Lanka's prolonged ethnic conflict over the past 25 years 'have the incurable habit of returning to war'. He argues that the puzzle of Sri Lanka's ethnic conflict between the government and the Liberation Tigers of Tamil Eelaam (LTTE) is embedded in 'the non-negotiability of the vital question of state power' and that 'protracted war has redefined the core issue of state power as one without negotiable options' (Uyangoda 2007, 1–2). He concludes that Sri Lanka's ethnic war is propelled by two contradictory and mutually exclusive state-formation agendas and war machines, one linked to the state 'reasserting majority nationalism' and one linked to the 'state-seeking minority nationalism spearheaded by the LTTE' (Uyangoda 2007, 45).

The post-2002 Ceasefire Agreement (CFA) period of 'No War/No Peace' allowed 'both parties to go through the motions of peace talks while exploring military options' (Centre for Policy Alternatives 2006, 11–12). In this period it seemed that the conflict 'found a new vitality toward remilitarization after four years of a ceasefire agreement, one year of peace negotiations, and the unprecedented humanitarian disaster of the tsunami which struck Sinhalese, Tamil and Muslim communities in almost equal measure' (Uyangoda 2007, 1). The tsunami struck areas under the control of both the government and the LTTE; however, initiatives for a joint post-tsunami administrative mechanism failed to create political reform through a process of engagement for humanitarian reconstruction and development (Uyangoda 2007, 24).

The February 2002 CFA between the government and the LTTE completely collapsed in January 2006. In January 2008, the government announced its withdrawal from the Norwegian-brokered ceasefire, formalizing a commitment to a military path (ICG 2008). It is estimated that some 5000 people (troops and civilians) have been killed since the beginning of 2006 (ICG 2008, 9). The situation deteriorated into open war in 2006, particularly in the east. The intensification of hostilities led to a serious humanitarian crisis, violations of human rights, and renewed displacement of over 290,000 IDPs (ICG 2007a, 7–8). Over 15,000 refugees left for South India (Centre for Policy Alternatives 2007, 4), adding to the existing caseload of over 60,000 in Tamil Nadu. In September 2007, the total number of IDPs stood at 503,000 (UNHCR 2007, 237). Armed conflict and human rights violations have led to displacement of all communities in Sri Lanka, both in LTTE (termed 'uncleared' areas by the government) and government-controlled areas (including 'cleared' areas).

The displacement crisis in Sri Lanka has fluctuated with the vicissitudes of war over 25 years, generating between 500,000 and one million IDPs of the total population of some 20 million. Although the war has displaced people from all three communities, it is estimated that Tamils make up *c*.80% of all IDPs in Sri Lanka, and in the late 1990s the diaspora of Sri Lankan Tamils in the west was estimated to be *c*.800,000 people (Schrijvers 1999, 309). According to the United Nations High Commissioner for Refugees (UNHCR), IDPs in the Sri Lankan context have three theoretical options for durable solutions: (1) returning to their place of origin (if and when it is safe); (2) local integration into the host community; or (3) resettlement elsewhere in the country.[3] However, 'durable solutions', including sustainable return, are problematic in an intractable conflict. Many who had been returned to their original areas had to flee again due to renewed conflict and insecurity. Return is thus always plagued with the risk of renewed displacement; hence, return is hardly 'durable'.

## Humanitarian protection

Current discussions on protracted internal displacement situations (Brookings-Bern 2007a) suggest a definition focused on the absence or failure of solutions rather than the duration or scale of displacement. The standard definition of a protracted refugee situation used by the UNHCR

(2000, 106), as 'one in which refugees find themselves in a long-lasting and intractable state of limbo' and which draw on the basic measure of refugee populations of 25,000 and greater who have been exiled for five years or more, does not resonate in most protracted IDP contexts. In the course of this research it became apparent that multiple cycles of displacement in Trincomalee compound the protracted *condition* of insecurity and belies a narrower or static definition of protracted displacement. This section outlines core components in our approach to humanitarian protection.

In sum, protection is about 'seeking to assure the safety of civilians from acute harm' (O'Callaghan & Pantuliano 2007, 1). Humanitarian protection means engaging directly with an understanding of the behaviour of actors and the dynamics of a conflict. As Mahony (2006) suggests, protection relies on a precise analysis of the specific situation. Sri Lanka demonstrates many of the archetypal characteristics of a difficult and unaccommodating protection environment: high levels of insecurity for humanitarian workers, a politicized humanitarian agenda and unpredictable levels of humanitarian access. Exploring operational perspectives on the challenges of humanitarian protection in politicized and militarized environments is necessary to ensure that research on protracted IDP situations reflects the ground realities.

Sri Lanka is an arena in which the international normative approach to internal displacement as set forth in the Guiding Principles on Internal Displacement is violated throughout all phases of displacement. Displacement is used as a tool or even a method of combat by both warring parties. Humanitarian agencies confront the tensions between agencies goals, the Sri Lankan government and non-state armed actors' goals and civilians' own perceptions of their predicament. Hence, approaches to displacement, protection, and 'durable solutions' need to be understood in relation to the political-military contours of the conflict. We draw on approaches that conceptualize humanitarian protection focused on 'civilian security' (Darcy 2007). The objective of humanitarian protection strategies and activities 'is to minimize threats of violence, coercion and deprivation, as well as enhancing opportunities to obtain security. The achievement of civilian security, or at least the reduction in insecurity, thus lies at the heart of protection' (Pantuliano & O'Callaghan 2006, 6). Protection encompasses physical and legal protection; primary responsibility for protection rests with the government, though international humanitarian law (IHL) imposes duties on all the parties to the conflict.

Protection and security for civilians involve three different actors: the authorities (governments, authorities and other bodies in control of a given territory, including armed groups and international forces); affected individuals and communities; and humanitarian agencies. The ICRC (International Committee of the Red Cross)-devised egg model covering responsive, remedial and environment-building spheres of action) emphasizes diversity and cooperation in the protection system, and how agencies can complement one another in their work with authorities, people at risk, civil society and each other.[4] As Slim & Bonwick (2005, 44–45) note, programming diversity can be a 'protection multiplier':

> A concerted effort by all agencies to use their different mandates, expertise, resources and networks to meet commonly identified protection needs and desired outcomes for threatened populations can dramatically increase the likelihood of protection being realized. Complementarity does not mean every agency doing the same thing. Instead, it involves each agency doing what it does best and what it is best placed to do.

The limits of humanitarian agencies as protection actors are widely recognized, as noted by the ICRC: '[h]umanitarian endeavors ... can never be a substitute for political action' (Aeschlimann 2005, 25). Humanitarian agencies can, however, support affected populations to avoid and resist threats through the following:

- Humanitarian presence: witnessing, accompaniment and monitoring, consciously used to inhibit abusive behaviour (Slim & Bonwick 2005, 91). The Centre for Humanitarian Dialogue's 'Proactive Presence' guide identifies three ways in which field presence contributes to the protection of civilians: deterrence (constraining abusers), encouragement (supporting people to protect themselves) and influence (supporting reformers and changing societal attitudes) (Mahony 2006).
- Protection-focused response and programming (assessing, designing, implementing, and monitoring programmes) takes two main forms: first, activities or interventions aimed at specific protection objectives or outcomes (specialist, dedicated protection programming); and, second, 'mainstreaming' or integrating protection into humanitarian programmes in order to minimize civilian risk (protection-sensitive assistance to build a protective environment). The Humanitarian Policy Group (HPG) identifies a third level for incorporating protection into humanitarian response – 'protective action', or activities that have both assistance and protection objectives or that are a means by which to address protection problems through assistance (O'Callaghan & Pantuliano 2007, 3).
- Protection through advocacy – influencing responsible authorities (government, as well as de facto authorities) to ensure respect for norms, rights and duties of international law.

In the next section, we contextualize the protection landscape of Trincomalee and the nature of human insecurity; then, in the following part, we consider in some detail field perspectives on the challenges of humanitarian protection, lessons learnt from operating in a protracted conflict setting, and creative strategies to negotiate these challenges.

## Protracted conflict in Trincomalee

At face value, the scale and frequency of displacement in Trincomalee, situated in the northern most part of the Eastern Province, has been comparatively less than that which occurred in northern Sri Lanka.[5] However, the Eastern Province has been an unstable and continually contested area throughout the war and is considered pivotal to Sri Lanka's future. The district is of strategic importance to the Sri Lankan government and the LTTE. Trincomalee harbour is the world's second deepest natural harbour, of geo-political and economic importance to both parties. It is also of particular nationalistic importance to the LTTE, which envisages Trincomalee as the capital of Tamil Eelaam. The strategic importance of the district bears significant implications for the insecurity of civilians in a context of unrelenting conflict.

Over the past 25 years, Trincomalee has experienced a complex process of 'mixing' and 'unmixing' of people (Rajasingham-Senanayake 2002, 60). The multi-ethnic composition of Trincomalee is unique in Sri Lanka. Unlike the mono-ethnic northern part of the country, today, Trincomalee has a fairly equal proportion of Muslims, Sinhalese and Tamils.[6] As McGilvray (2001) shows, Trincomalee has experienced decades of demographic manipulation. More recently, the introduction of 'buffer zones' in 2005 (following the tsunami), a 'Special Economic Zone' in 2007 and also the High Security Zones, has spawned allegations of Sinhalese 'colonization' in government land use, development and settlement policy (Centre for Policy Alternatives 2007; Hyndman 2007; Rampton 2007).

Trincomalee has experienced a high level of political violence that is expressed in suspicion and mistrust between different ethnic groups of the population (Foundation for Coexistence 2004). The defection of Colonel Karuna from the LTTE in March 2004[7] generated intra-ethnic tensions which were felt at community level, and the renewed war in 2006 exacerbated tensions between Eastern Tamils and Muslims (ICG 2007b). The politicization of ethnicity and violence has polarized communities and produced a volatility that contains features unique to the east of the country (Philipson & Thangarajah 2005). Causes and solutions are localized in the eastern context and within its sub-regions; protection problems therefore contain specific geo-political and community sub-dynamics and require locally contextualized understanding.

Trincomalee District has undergone waves of displacement since the 1980s, and again in 2006 it became a main site of displacement in Sri Lanka. In April 2006, bombing in Trincomalee town triggered communal violence and displacement; later in April mass displacement in LTTE-controlled areas in Muttur and Eachchilampattu DS Divisions resulted from government military operations, especially around Sampur. In August 2006, the government launched a military offensive in response to the LTTE shutting the sluice gates in the Serunwara area. The LTTE launched a counteroffensive, seizing control over parts of Muttur and provoking massive retaliation by government forces. The civilian population of Muttur and its locality fled *en mass* within the district and into neighbouring Batticaloa (IASC 2007b, 39–52).

In July 2007, the government officially declared the Eastern Province under its control ('cleared' of the LTTE) and ready for development, but influential agencies (such as the ICRC and the Sri Lanka Monitoring Mission (SLMM) still consider the region a conflict area. Political underpinnings of Sri Lanka's 'conflict-sustaining trajectory' (Uyangoda 2007, viii) ensure that peace remains a distant prospect overall. In the context of an enduring capacity for war, people have experienced repeated displacement for more than two decades. Experiences of multiple cycles of displacement during this protracted period have perpetuated a state of uncertainty, impermanency and insecurity (interviews, 2007). Attention therefore needs to be directed to understanding the sources of civilian insecurity and the most effective responses available to support protection and human security at community level.

The conflict-related displacement landscape is marked by forced displacement, restrictions on freedom of movement, and forced returns and relocations (IASC 2007b). Protection concerns also include extreme human rights abuses such as disappearances, abductions for ransom, extra-judicial killings, forced recruitment (including children), arrests and detentions, 'round-ups' to identify suspect individuals by 'masked men', extortion, and wider insecurities connected to the loss of livelihoods and the impacts of militarization. Risks, threats and vulnerabilities are pervasive and intricately part of every aspect of people's lives: 'the environment is such that we [agencies] can't avoid protection [in our work]' (interview, 2007).

## *'We don't dare to hope' – human insecurity*

Civilians are not only collateral damage in this war, but are often deliberately targeted as part of the military strategy of parties to the conflict (state, armed non-state, and paramilitary actors). The scale and intensity of military operations in 2006, including aerial bombardments and shelling (use of 'multi-barrels') targeting civilian areas, have compounded returnees' sense of insecurity in this phase.

The shift in military-political balance to the government in return areas formerly controlled by the LTTE has also shaped civilians' perceptions of their ability to flee: 'we are surrounded in this place' (returnee family, interview, 2007). Also, as noted by the ICRC (interview, 2007), military structures (such as army camps and buffer zones) in Sri Lanka are in close proximity to civilians. The militarization of return areas as well as past experiences of retaliatory attacks undermines prospects for sustainable return: 'when we get back [return] the military will look at us ... If something happens the military will shoot.' (returnee families, interview, 2007). People's sense of security is also influenced by specific local factors such as the behaviour of local military authorities, the proximity of their homes to contested areas (jungle, coast, etc.), and events occurring in neighbouring villages.

People expressed a pervasive sense of insecurity in their lives and prospects for their future: 'we don't have hope [for the security of our future] at all. We don't dare to develop hope' (returnee family, interview, 2007). Active hostilities may have ceased for the time being and people have returned home, but they do not feel secure. They are not sure when the conflict might reignite and when they will need to run again. Thus, an imminent sense of a renewed conflict marks their return.

Those who have returned preferred to be back in their own villages but noted that political divisions and mistrust are being played out at community level: 'before, one or two [parties to the conflict] were there. Now there is one, two, three, or four'(returnee family, interview, 2007). Political factionalism is producing and reproducing threats and vulnerabilities. Political factions in the Tamil community, for example, have led to a breakdown in the collective coping mechanisms used by communities in the past: 'We can't even trust the next door neighbour [now]. We don't know who's coming or for what reason' (returnee family, interview, 2007). Political violence is played out at the personal and family level, such as in the case of abductions carried out for personal vengeance (CBOs, interviews, 2007). Returnees following the 2006 conflict expressed renewed and intensified forms of suspicion, mistrust and divisions resulting from their experiences of displacement and enduring political violence (returnee families, interviews, 2007). One local commentator described the political violence as the most serious she had known. The next section explores implications for agencies working in this conflict environment, with multiple cycles of displacement and continuing insecurity.

## Protection challenges for humanitarian agencies

In eastern Sri Lanka, humanitarian protection confronts the highly militarized, politicized arena and/or the denial of access in some areas. It is an environment in which all political actors (state, non-state, paramilitary) are perpetrators of violence and insecurity, and international principles and rules for the protection of conflict-affected civilians and IDPs are not respected. Agencies therefore confront 'strategic risks' (Slim & Bonwick 2005, 46–47) and operational dilemmas in their protection work. In general, agencies working in the context of *internal displacement* are often constrained by the policies, actions or inactions of governments. Governments have the primary responsibility for the protection of IDPs and affected civilians, but too frequently they fail to protect and/or are themselves sources of insecurity. International Humanitarian Law obliges all parties to distinguish between the civilian population and combatants at all times. In Sri Lanka, the failure of the government to protect civilians constrains and frustrates humanitarian actors working in the midst of the conflict-affected environment.

The key operational challenges for humanitarian agencies relate to negotiating the politically sensitive relationships with the government, dealing with a militarized and deeply divided environment, and internal factors influencing how agencies approach their work. The sections below outline these challenges, but first several introductory points need to be established. 'Agencies' is the vernacular catch-all term,[8] but for the purposes of disaggregating agencies' perspectives we have broadly categorized agencies as: (1) international agencies encompassing the UN agencies, the International Committee for the Red Cross (ICRC),[9] and international non-governmental organizations (international agencies include protection mandated agencies charged by states to lead on a particular aspect of humanitarian protection, and also non-protection mandated agencies, i.e. humanitarian NGOs offering humanitarian action in support of protection, and that recognize the importance of linking protection and assistance to people affected by conflict); and (2) local agencies encompassing local non-governmental organizations (NGOs are registered through a national secretariat in Sri Lanka) and community-based organizations.

### *Working with the Government*

International agencies are primarily concerned with the *conditions* under which the war is played out, although some agencies reflected on how the provision of humanitarian assistance has been used to propagate political agendas contained within the government's securitized policy narrative. Some international agencies have become concerned about the government's exploitation of the 'No War/No Peace' period for the militarization of 'liberated' regions and the securitization of development. Both governmental objectives were pursued by instigating rapid returns that did not meet internationally agreed benchmarks for conditions and processes of return.

Agencies risk instrumentalization of assistance by government for political purposes, which is a process by which assistance is used as a tool for different political ends by the government. Working with the government can present agencies with the dilemma of how to carry out their work without the 'risk of implementing a plan that is not a humanitarian plan' (international agency, interview, 2007). Agencies identified the need to avoid the risk of incorporating assistance into a plan that legitimizes forced return or that contributes to neglect or abuse of a population. One area is the instrumentalization of return (termed 'resettlement' by the Sri Lankan Government) by the military for political purposes. For example, according to some international agencies rapid (mass) return occurred in order to consolidate the government's military campaign by bringing back civilians into an area at risk of incursions by the LTTE. Rather than being an agenda for sustainable return, 'to the government, return means surveillance and [military] security' (international agency, interview, 2007). The return of people enables the government to more tightly control and scrutinize the population through screening and identification of people formerly living in LTTE areas.[10] Agencies have been faced with the problem of how to engage with militarized return as the government did not provide a resettlement plan, and initially agencies were unable to access return areas in Eachchilampattu DS Division, and also the town of Vakarai (in the Batticaloa District of the Eastern Province). As another international agency head explained, 'on the one hand we are working with the state as a real partner but at the end of the day when you need real commitment they have their own political and military agenda' (international agency, interview, 2007). Opposing narratives are expressed in the contradiction between the international normative framework to protect IDPs and the policy narrative of the Sri Lankan government which envelopes protection and solutions to displacement with a security agenda.

Agencies noted political interference in their humanitarian work plans, described by one observer as 'à la carte humanitarianism' (international agency, interview, 2007), or a situation in which the government picks and chooses activities according to its own agenda. For example, agencies' access has improved in resettlement areas in Eachchilampattu, where the government is keen to reconstruct the area (for an overview of securitized development in eastern Sri Lanka see Rampton 2007). Agency staff noted that the high regulation of agencies by the government, which has gradually increased in response to the proliferation of operational agencies in post-tsunami Sri Lanka, impinges on their independence. For example, the government has issued 'guidelines for NGOs/INGOs in their Dos and Don'ts which stipulates that agencies should "not waste time and resources" on "awareness programs"' (international agency, interview, 2007).

Agencies are contending with a politically confronting environment where protection work is considered highly sensitive because it is negatively perceived as challenging the agendas of political and military actors. International and especially local agencies noted that 'you can't be outspoken on protection'. It is not possible to use the 'P-word' (protection) openly. As one veteran of a CBO (interview, 2007) responded: 'Your subject of protection is a dangerous one.' Protection work can risk jeopardizing agencies' relationships with the authorities due to the uncomfortable exposure of the government's failure to protect its people and the political reality underpinning the causes of protection problems. Agencies in this environment thus need to consider the political consequences of undertaking protection work and risk of 'protection backlash' (provoking the ire of the authorities) (agencies, interview, 2007).

The sensitivities of the political environment mean that some agencies compromise their principles (agencies, interview, 2007) in order to remain operational, i.e. their principles referring to adhering to agencies' mandates and to supporting and promoting the Guiding Principles on Internal Displacement. Agencies find themselves balancing the need to stay operational with negotiating the political agenda of the government. Further, international agencies do not take a strong principled position on situations from the outset of engagement, which leads to erosion of humanitarian principles and a breakdown of collective action of agencies. Despite the endorsement of the *Saving Lives Together* framework (IASC 2006), prepared by a sub-working group of the IASC Task Force on Collaborative Approaches to Security led by InterAction and UNICEF in November 2006, there remains limited collective action and solidarity (international agency, interview, 2007). Agencies will lobby for parties to adhere to humanitarian principles but proceed to work within the prevailing conditions.[11] Even advocacy in the form of quiet diplomacy with government authorities may amount to 'lipstick service to our principles' (international agency, interview, 2007). Failure of principled action inadvertently contributes to legitimizing actions that contravene the Guiding Principles on Internal Displacement.

The humanitarian response to the natural disaster of the tsunami that struck Sinhalese, Tamil and Muslim communities on the coast of Sri Lanka in December 2004 exacerbated the breakdown of collective action. The influx of relief funds in post-tsunami Sri Lanka created 'competitive humanitarianism' (Stirrat 2006). The high level of competition between international agencies to access donor funds and competitively bid to implement projects with the Sri Lankan Government led to a breakdown in collective action between international agencies and an increase in 'marking territory'. Further, the tsunami response led to another inundation in the form of international agencies that were not necessarily prepared for or skilled in working in a conflict-affected environment, resulting in some international agencies not taking a strong principled position on situations from the outset of engagement.

Coordination with the government has been problematic. Agencies must deal with a centralized administration dominated by military interests. To give an example, the most powerful civil administrator in Trincomalee District, the Government Agent (GA), was formerly a major general. Agencies are constrained in their planning by ad hoc and inconsistent government decision-making processes, including a lack of consultation regarding a consolidated resettlement and rehabilitation plan. For many humanitarian agencies coordination with civil administrators charged with military objectives is deeply problematic as it can result in the compromise of principles of independence. Further, agencies contend with different models of the state in different regions and sub-regions. Civil administration decisions are connected to the larger military campaigns of the Sri Lankan Army and the LTTE. For example, negotiations at the local level of the District Secretary (DS) can boomerang at the level of the Government Agent (GA) as military priorities take precedence (international agency, interview, 2007). This endemic feature of the protection environment creates certain requirements in agencies' initial engagement in order to understand power structures and to set objectives.

Agencies noted the government may perceive their work as biased in favour of one population (notably the Tamils, whom have been disproportionately affected by displacement). This apparent bias can result in the erosion of perceived neutrality in the eyes of the government: 'to assist people is a political statement' (international agency, interview, 2007). There are entrenched perceptions held by Sinhalese nationalists, the government and armed forces of NGOs as 'criminals', spoilers, and wrongdoers (international agency, interview, 2007). For humanitarian agencies seeking to engage with civilians displaced by war, Tamils make up *c.*80% of all IDPs in Sri Lanka. Agencies therefore confront the dilemma of maintaining unbiased engagement whilst ensuring a commitment to engaging with the most affected populations. Agencies must remain vigilant about how the authorities perceive their work: 'For us, when we had meetings with the Government Agent we had to list the Sinhalese villages in which we worked first and then the Tamil areas so as not to raise a red flag' (international agency, interview, 2007). Furthermore, as the majority of the displaced populations are Tamil, agencies require Tamil staff to be able to communicate with affected communities. Most Sri Lankans are not bilingual in Tamil and Sinhala, although. Muslims are Tamil speaking. However, ethnicized relations largely necessitate agencies to hire Tamils to work with Tamil communities to ensure the safety of their staff and to enable more accurate information gathering. The high proportion of Tamil staff in international agencies has compounded perceptions of bias by the authorities and is used as evidence to reinforce the suspicion that INGOs are supporting the LTTE.

Protection work also exposes local staff to risks of threats from various armed actors (such as the Tameleela Makkal Viduthalai Puligal, TMVP, the political arm of the Karuna faction) that perceive their involvement as biased or interfering.[12] There is reluctance on the part of local agencies to work on protection-related issues due to concern for the security of their staff. Similarly, international agencies have to weigh decisions against the potential cost to staff, programmes and communities. A national 'culture of impunity' (interviews, 2007) and the lack of outcome or noticeable change from humanitarian reporting leads to a reluctance to report incidents, as deliverance is weighed against the risk of reporting. The breakdown in trust within and between communities constrains specific case reporting to local agencies, as the impartiality of local agencies is not guaranteed.

### *Wider contours of the operational environment*

Agencies are limited in their capability to contribute to the protection and security of conflict-affected IDPs and civilians due to features inherent in the militarized environment that impact access, presence, and monitoring and reporting. Agencies are constrained by limited access determined by government as well as their own security guidelines (which, for example, require them to return to base by sunset). Within UN agencies, for example, it is a requirement that local staff be accompanied by international counterparts when conducting fieldwork, thus placing a strain on resources available for field presence. Local agencies and communities are calling for presence at the field level as a preventative mechanism to deter perpetrators as well for reporting leverage and empowerment for communities, yet international agencies are constrained by the dangers and risks of the operational environment (in addition to their own mandates) to maintain humanitarian presence in affected areas. Many human rights violations occur at night and this is an aspect of protection that is not addressed in the current responses by international agencies (with the exception of the work of Non-Violent Peace Force). In addition, for local agencies and staff, access and security conditions are also garmented in fear and mistrust between ethnic groups.

Precise reporting and referrals are also limited as the process presents risks for victims and witnesses. People fear speaking out and reporting on protection incidents due to the risks of retaliation by parties identified as perpetrators. Community-based organizations (CBOs), too, are reluctant to report protection incidents because of lack of witness protection and risk of revenge and intimidation of their own staff. This limits specific case reporting. Humanitarian agencies have put together a referral mechanism for protection cases in Trincomalee which details how agencies can refer protection cases that they come across in their daily work, and explains the mandates of protection agencies. However, in practice, timely action necessary for effective reporting and follow-up of cases is not always delivered. Agencies are often unable to verify information due to communities' fears and/or access to cooperative authorities to cross-check details of cases. Further, some local NGOs and CBOs are insufficiently informed of international agencies' mandates, scope of action, and the establishment of a referral mechanism. Lack of communication and clarity between international and local humanitarian agencies also hampers local agencies from utilizing the referral system.

### *Internal views*

Institutional challenges, such as diverse attitudes and awareness of protection, rigid views of institutional mandates, and lack of experienced personnel pose challenges for agencies. Some international agencies are blinded by their own mandates – they may be 'pushing their own agenda, [which] means missing the bigger picture'; or they may concentrate on 'the small things we have in front of us' (international agency, interview, 2007) rather than take a longer term view.

Agencies have different understandings of humanitarian protection. 'Protection' in the context of Trincomalee is largely understood in terms of basic physical security and the need to ensure safety. Particularly for local agencies, the protection agenda is limited to a concern with physical security, namely 'hardcore' protection issues of abductions and killings. For local agencies there is uncertainty about the scope of how internationals protect civilians, and what the international protection agenda is. Some non-mandated protection agencies work well with mandated agencies (for example, utilizing the referral mechanism) and approach their work with a protection-sensitive orientation. Other non-mandated agencies understand protection as requiring a specialist mandate: 'we are not a protection agency' (international agency, interview, 2007). Rather than working with a wider understanding of protection-sensitive response and programming, protection is seen as the responsibility of particular specialist humanitarian agencies charged with protection mandates or offering expertise in protection.

Most non-mandated agencies do not have systematic approaches to protection-sensitive programming but depend on local staff to identify protection issues in their programme design and also the varying capacity of international staff to consider security issues in programme development. Largely, non-mandated agencies did not view technical interventions (e.g. water and sanitation projects) as an opportunity to provide protection nor did they regard protection as intrinsic to assistance. Uncooperative relationships between the government and NGOs have hampered NGOs' influence. Non-mandated agencies have limited bargaining power and frequently no political leverage or relevance to the authorities. Non-mandated agencies rely heavily on protection-mandated agencies (in the form of the

ICRC, the UN and the former Sri Lanka Monitoring Mission) to access government authorities and to advocate on issues. Non-mandated agencies are reliant on the lead agency (the UNHCR) to take a principled stance as an authoritative voice and representative of humanitarian interests to ensure common protection goals.

Agencies at the field level noted internal constraints of donor clocks (such as benchmark requirements in the project cycle), headquarter processes and demands as constraining flexibility, creativity and responsiveness. The pulsating nature of the war (alternating flares between the north and east coast) means agencies' resources and high-capacity personnel are often reallocated to the new emergency areas, reducing investment in and commitment to the ongoing protection crisis in comparatively 'stabilized' areas. In some cases, non-mandated protection agencies expressed a lack of headquarters' support for protection activities due to the perception that protection advocacy is too politically risky in terms of jeopardizing established relationships at the Colombo level. Mandated agencies noted that they are reliant on Colombo to be a vocal advocate of their recommendations.

Cooperation and coordination between international agencies depends heavily on personalities. Good collaboration relies not on the organization but on particular persons within an organization. Protection agendas are driven more by personalities and the confidence and knowledge of individuals on the ground than by an institutional system-wide agenda. Ensuring that agencies sustain coordinated engagement is an 'ongoing battle' (international agency, interview, 2007). Coordination and engagement with local agencies has been limited due to constraints of a weak civil society, security risks to nationals posed in a politicized environment, and the limited investment of international agencies in utilization and building of local capacity.

The tsunami response also led to the creation of many local NGOs and CBOs that did not exist before the tsunami but were formed to access international donor funds (interviews, Trincomalee, 2007). The proliferation of local NGOs and CBOs led to suspicion amongst INGOs of partnering with these local organizations as patterns emerged of newly formed CBOs and/or local NGOs misusing funds and not fulfilling their project obligations. A post-tsunami residue of suspicion of local NGOs and/or CBOs and a lack of collaboration between INGOs and local agencies still exists today. Reluctance by international agencies to partner with local agencies limits collective action between local and international humanitarian agencies. Limited support for the mobilization of local agencies hampers local agencies from capitalizing on the influence and access to authorities that international agencies possess. For international agencies, partnership with local agencies, particularly CBOs, ensures greater access to the community. However, international agencies need to remain aware of the potential politicization of local agencies and weigh these opportunities and risks and then decide when, where and how to engage with local agencies (international agency, interview, 2007).

International agencies responding to the tsunami and development actors in general were not prepared or skilled for operating in a complex, conflict-affected environment of intractable war: 'here, there is a lack of interest in protection. People start thinking things are normal when they are not' (international agency, interview, 2007). As another agency put it, 'considering what's going on. There are not many agencies really getting it' (international agency, interview, 2007).

## New challenges, old lessons

Protection strategies to incorporate civilian risk and resist threats to the security of affected populations relate to all areas of protection work – presence, protection-sensitive programming, and advocacy. Responses encompass the immediate and long-term complementary spheres of responsive, remedial and environment-building action contained in the 'egg model' (Slim & Bonwick 2005, 42). Clearly, there are different agencies, with different organizational structures and mandates. However, it is useful to distil lessons from the various perspectives. The case of Trincomalee illustrates how the politicization of humanitarian protection by all parties to the conflict requires agencies to develop context-specific, creative approaches to their work. Overwhelmingly, agencies reflected on the need to develop humanitarian action that is informed by a political understanding of the context and consistently take a principled and protection-sensitive approach.

Agencies have been limited in the extent to which they can counter the risk of instrumentalizing assistance for political purposes. However, there has been a concerted effort to build productive relationships with the high-level administrators (especially the powerful Government Agent) which has allowed agencies greater negotiation space and the possibility for influence. For example, recent efforts to consult with the Government Agent on humanitarian plans prior to public coordination meetings between government and humanitarian agencies have improved government-agency relationships.

For non-mandated agencies, it is particularly important to maintain a low profile in how they perform their protection work. Agencies have had to avoid publicly using the 'protection' label, but rather focus on building a protection-sensitive orientation into their work. In this environment, protection is as much an orientation as a set of activities. Long-term engagement with Sri Lanka has enabled agencies to decipher and utilize non-controversial entry points (such as activities focused on income generation, women and children) that are perceived as less politically confronting. As the local conflict dynamics are in flux, agencies need to remain vigilant regarding how communities and the authorities perceive their protection work.

For the protection-mandated agencies, it is important to work transparently and confidently according to the principles of the agency and to ensure the government is held accountable for its responsibilities. Today, a lack of communities' confidence and trust in the government has resulted in reliance on international agencies (such as UNICEF and ICRC) rather than the state (including the police), which in turn contributes to the stagnation of the existing system. It is therefore important to build government accountability and capacity.

The advice given by one international agency during an interview in 2007 was 'Stick to principles!' Agencies have principles and privileges and they need to use them. An important lesson revealed was that agencies should maintain principled action from the outset of their engagement with the authorities. As this article has shown, the failure of principled action inadvertently contributes to legitimizing actions that contravene the Guiding Principles on Internal Displacement and the breakdown of collective action by agencies. This applies to the larger and UN protection-mandated agencies in particular, as these are agencies with access and influence. As one key respondent noted, taking a principled stand should mean that protection-mandated agencies 'don't become intimidated [by the authorities] and do it anyway' (international agency, interview, 2007). In other words, agencies should be consistent in their application of principles. In addition, maintaining principled action also ensures predictability of international agencies' scope of action for local and national agencies that rely heavily on international agencies to take a lead role in setting and establishing the humanitarian protection agenda. Protection-mandated agencies play a quasi protection role for local and national agencies, particularly in forging and establishing the humanitarian protection agenda with the government. A 'protection chain' between international, national and local agencies allows for a balance between protection and participation of local and national agencies in an insecure environment.

Agencies have addressed the government's perceptions of bias by actively communicating with the relevant authorities to demonstrate commitment to a balanced approach with all three communities and to create visibility to their work with all three affected communities. Agencies also are engaging with IDPs in conjunction with host communities to ensure that agencies' do not fuel mistrust and resistance between and among communities.

However, effective communication and coordination is essential, since because a considerable number of agencies do not identify themselves as humanitarian protection actors, coordination efforts struggle to garner the diversity of actors that can contribute to protection outcomes. If agencies are well-coordinated, this enables a 'protection multiplier' effect, bringing together agencies with their different mandates, expertise, resources, and networks to meet commonly identified protection needs. Further, non-mandated international agencies are utilizing the Trincomalee referral system for protection cases (reporting of abductions, killings, etc.) to benefit from the access and influence that mandated agencies possess.

Strengthening community protection capacity is particularly important in an environment in which political divisions are manifesting at the community level and are breaking down traditional community and/or collective coping mechanisms and also the capacity of civilians to protect themselves. International agencies are realizing the inherent value of engaging at community level and utilizing community capital as an entry point for engagement.

Local agencies are calling for international agencies to support the mobilization and capacity building of local agencies, and to pursue long-term environment building approaches. In the current phase of return and rehabilitation, local agencies see an urgent need for international agencies to support local agencies to create attitudinal change in divided communities and build trust. Local agencies highlighted the challenges of engaging in a context of 'generalized' violence and understanding the meaning of violence in a Sri Lankan context, '[international] agencies don't see the Sinhala-Tamil dynamic as reproducing violence; they do not understand spirals of personalized violence' (international agency, interview, 2007). Insights from local agencies suggest that protection-sensitive programming should closely consider linkages with peace building in a context of protracted conflict.

There is a vital space for protection and human security at a very local level. Building community resilience to safeguard against the breakdown in community protection strategies is necessary in the context of political factionalism and fractured communities. Creating shared mechanisms and interests between and within communities should be incorporated into approaches to build opportunities for resilience in divided communities. Agencies have the opportunity to incorporate preventative action and contingency planning in the reconstruction of return areas, for example, creating shared community facilities to build trust. Further, the 'recovery and rebuilding' phase of the conflict provides an inlet for packaging programmes in economic objectives, such as creating joint or mixed livelihood programmes between different communities as a protection mechanism to support the prevention of renewed inter-ethnic and intra-ethnic tensions.

Protection needs to be analysed at a village level, and agencies are realizing that taking a longer term commitment to particular villages rather than pursuing available spaces (such as priority areas for the government) for intervention enables agencies to more effectively understand the threats, vulnerabilities and capacities of communities to build locally resonant approaches. Partnership with CBOs ensures genuine commitment to the community: 'they [CBOs] are not part of the local NGO industry game – they will not leave' (local agency, interview, 2007). However, international agencies need to remain aware of the potential politicization of local agencies in the ethnicized and politically fractured environment. For international agencies, identifying impartial partners, ensuring the confidentiality of individual protection cases whilst working with local or national partners and managing the security of local and national partners creates disincentives for establishing partnerships with local humanitarian actors.

Often perceived success can be divorced from the realities of engaging with civilian insecurities. Agency pressure to demonstrate success to donors prevents agencies from approaching protection as a *process* – one in which the situation transforms itself, demanding new responses. Longer term engagement is necessary to address entrenched features of the protection environment to support attitudinal change: 'it is difficult to influence behaviour; it takes time' (local agency, interview, 2007). A commitment to capacity building of local authorities and humanitarian agencies is essential. As one international agency reflected, '[c]onsider us [international agencies] short-term, consider them [local agencies] long-term' (international agency, interview, 2007). Local agencies want stronger and genuine partnerships: 'we expect mutual partners; not only funds, not only an outsider observing everything' (local agency, interview, 2007). Local agencies reported that international agencies must communicate clearly their role and limitations in order to ensure accurate expectations and facilitate greater collaboration between international and local protection actors.

## Concluding remarks

Since 1983, displacement in Sri Lanka has been embedded in intractable conflict which has produced an ongoing condition of insecurity. Many IDPs have experienced prolonged, repeated and ongoing cycles of displacement, whilst some IDPs have stayed in the same place for more than 10 years. Durable solutions, especially sustainable return, are problematic in this context. Many who have returned to their original areas have had to flee again due to renewed conflict and a precarious security and protection environment. The failure of lasting solutions in the absence of a transformative peace process requires humanitarian agencies to understand the discreet phases of renewed displacement against a landscape of the longer term, unresolved nature of the conflict that continues to (re)-produce insecurities.

Humanitarian agencies must contend with how to approach internal displacement in difficult and insecure environments of protracted crises. Humanitarian agencies seeking to assist IDPs have to grapple with the policies, actions or inactions of government that has primary responsibility for protection of IDPs but often fails to protect or is a source of insecurity. There are unique challenges for humanitarian agencies in protecting IDPs in contexts characterized by the politicization of humanitarian protection, fluctuating access and insecure environments for aid workers. These operational realities need to be taken into account when developing approaches to protracted conflict and IDP situations. While agencies cannot be a substitute for political action to resolve protracted displacement, they can pay particular attention to how they approach their work that is informed with an understanding of the threats, risks and vulnerabilities confronting conflict-affected communities. This requires analysing the security and political dimensions of protection problems within the operational environment. The underlying question is *how* international actors, donors and agencies work with local actors to determine their involvement on a continuing basis. Pragmatic and locally contextualized humanitarian approaches, that prioritize the perceptions of the internally displaced to the kind of threats, risks and vulnerabilities they confront, can allow for changes over time that impact dynamics within and across different communities.

## Notes

1 Sir John Holmes, UN Secretary General for Humanitarian Affairs, described the gunning down on 6 August 2006 of 17 Action Contre la Faime (ACF) employees in Muttur town, Trincomalee District, as the 'single worst crime committed against humanitarian workers in recent history' (IRIN 2007).

2 The results of this research are reproduced with the publishers' permission from 'Conflict-induced internal displacement: Challenges and dilemmas of protection for humanitarian agencies in Trincomalee', in *Protracted Displacement in Asia: No Place to Call Home* (Ashgate, Aldershot, 2008), pp. 107–131. Copyright © 2008, and are published in the research report *Protracted Conflict: Protection Challenges for Humanitarian Agencies*, Austcare 2008, available at: http://www.austcare200X.org.au/media/56973/arcsrilankabp-lowres.pdf (accessed April 2008).

3 UNHCR. Internally Displaced Person (IDPs) Working Group Workplan and Supporting Documents, Colombo, May 2006. Unpublished.

4 The 'egg model' spans three interdependent, complementary spheres of action (Slim & Bonwick 2005, 42), responsive (immediate or urgent activities aimed to stop, prevent or alleviate the immediate and/or worst effects of abuses); remedial (restorative action to assist people to recover while they live with the effects of abuse); and environment-building (supporting political, social, cultural, and institutional norms that prevent or limit violations or abuse).

5 The conflict zone in Sri Lanka is customarily divided into the three areas: the Vanni – mostly controlled by the LTTE; Jaffna, and the border areas south of Vanni – areas which the government has regained and which it calls 'cleared' areas; and the contested areas of Trincomalee, Batticaloa and Ampara in the East which have been declared 'cleared' by the government since July 2007.

6 A reason for the unique ethnic composition was the state-sponsored irrigation and Sinhalese settlement schemes that began in the 1940s (Peebles 1990), a process that radically altered the political demographics of Trincomalee. Previously, the east coast was a region composed of two historically interlinked, but contentious communities, the Tamils and the Muslims, both of which are Tamil-speaking and have a long history of conflict and co-existence (McGilvray 1999, 2001).

7 In 2007, the Tamil Makkal Viduthalai Puligal (TMVP) was formed as the political arm of the 'Karuna' faction. The group is 'now led by the ex-Colonel Karuna's former deputy, Pillayan, continues to rule Batticaloa and other parts of the east through terror and crime, with tacit police, military and Colombo approval. Still seen by the government and military as useful to block a Tiger re-emergence in the east, its reign of abductions, child recruitment, robberies and repression of dissent is extensively documented.' (ICG 2008, 12).

8 In order to protect the confidentiality and security of interviewees, we have not attributed quotes and statements directly to individuals or their organizations.

9 The ICRC's mission – defined in the 1949 Geneva Conventions and their 1977 Additional Protocols and in the 1986 Statutes of the International Movement of the Red Cross and Red Crescent – is 'to protect the lives and dignity of victims of war and internal violence and to provide them with assistance' and 'to prevent human suffering by promoting and strengthening humanitarian law and universal humanitarian principles.' Protection is therefore of primary concern to the International Committee of the Red Cross (ICRC). However, ICRC is under no obligation to coordinate with other agencies.

10 New, duplicate identification cards in Sinhala language have been issued to returnees from formerly LTTE-administered areas, labelling people with 'red letters'. Mobile phone coverage is blocked in the return areas of Eachchilampattu, further impinging on agencies' security, and constraining involvement in and approaches to the area. The Sri Lankan Army has established camps in return areas to create 'buffer zones'.

11 This has been seen, for instance, in a reticence to be associated with building alternative housing for returnees whose homes are occupied by the army. Whilst working towards humanitarian goals (such as the need to provide shelter), this action has the potential to assist in the consolidation of military objectives.

12 The TMVP is widely regarded as a proxy to the government. The strengthening of TMVP has created political and criminal obstacles for agencies to negotiate. The TMVP are considered less disciplined than the LTTE and are viewed by some as a criminal group rather than a political movement.

*Acknowledgements.* – We are grateful for the generosity and assistance provided to us in Trincomalee by Austcare's partners in the field. We wish to extend our appreciation to respondents in agencies working on the ground who kindly gave us their time and reflections. This research was supported by an Australian Research Council funded project between Griffith University, the Australian National University, and the humanitarian NGO Austcare.

# References

Aeschlimann, A. 2005. Protection of IDPs: An ICRC view. *Forced Migration Review* Supplement, October 2005.

Brookings-Bern Project on Internal Displacement. 2007a. *Expert Seminar on Protracted IDP Situations.* 21–22 June 2007. Geneva. http://www.brookings.edu/events/2007/ ~/media/Files/events/2007/0621_displacement/20070621_displacement.pdf (accessed December 2008).

Brookings Institution-University of Bern Project on Internal Displacement. 2007b. *When Displacement Ends: A Framework for Durable Solutions.* Brookings Institution – University of Bern Project on Internal Displacement, Washington DC.

Centre for Policy Alternatives. 2006. *War, Peace and Governance in Sri Lanka: Overview and Trends 2006.* Centre for Policy Alternatives, Colombo.

Centre for Policy Alternatives. 2007 *Policy Brief on Humanitarian Issues.* Centre for Policy Alternatives, Colombo.

Darcy, J. 2007. Political and humanitarian perspectives on the protection of civilians. Paper prepared for HPG Geneva Roundtable on Protection, 22 January 2007. Available at http://www.odi.org.uk/hpg/meetings/protection_conceptsnote.pdf (accessed 24 September 2007).

Foundation for Coexistence. 2004. *Ethnic Relations and Human Security Eastern Sri Lanka.* Asia Foundation,Washington DC.

Hyndman, J. 2007. The securitization of fear in post-tsunami Sri Lanka. *Annals of the Association of American Geographers* 97:2, 361–372.

IASC (Inter-Agency Standing Committee). 2006. *Saving Lives Together: A Framework for Improving Security Arrangements Among IGOs, NGOs and UN in the Field.* Prepared by a sub-working group of the IASC Task Force on Collaborative Approaches to Security (led by InterAction and UNICEF). http://www.icva.ch/doc00001990.pdf (accessed March 2007).

IASC (Inter-Agency Standing Committee). 2007a. *Benchmarks for Durable Solutions for Internally Displaced Persons.* Brookings Institution – University of Bern Project on Internal Displacement, Washington DC.

IASC (Inter-Agency Standing Committee). 2007b. *Conflict-Related Internal Displacement in Sri Lanka: A Study on Forced Displacement, Freedom of Movement, Return and Relocation.* April 2006 – April 2007 and Addendum covering April – July 2007. http://www.unhcr.org/protect/PROTECTION/482ae16b2.pdf (accessed 5 October 2007).

ICG (International Crisis Group). 2006. *Sri Lanka: The Failure of the Peace Process.* Crisis Group Asia Report No. 124. International Crisis Group, Brussels.

ICG (International Crisis Group). 2007a. *Sri Lanka's Human Rights Crisis.* Crisis Group Asia Report No. 135. International Crisis Group, Brussels.

ICG (International Crisis Group). 2007b. *Sri Lanka: Muslims Caught in the Crossfire.* Asia Report No. 134. International Crisis Group, Brussels.

ICG (International Crisis Group). 2008. *Sri Lanka's Return to War: Limiting the Damage.* Crisis Group Asia Report No. 146. International Crisis Group, Brussels.

IRIN. 2007. Sri Lanka: UN urges investigations of deaths and more security for aid workers. UN Office for the Coordination of Humanitarian Affairs. 7 August 2007. http://irinnews.org/PrintReport.aspx?ReportId = 73621 (accessed 7 December 2007).

Macrae, J. & Harmer, A. 2004. *Beyond the Continuum: An Overview of the Changing Role of Aid Policy in Protracted Crises.* Humanitarian Policy Group, Overseas Development Institute, London.

Mahoney, L. 2006. *Proactive Presence, Field Strategies for Civilian Protection.* Centre for Humanitarian Dialogue, Geneva.

McGilvray, D. 1999. Tamils and Muslims in the shadow of war: Schisms or continuity? Gamage, S. & Watson, I.B. (eds.) *Conflict and Community in Contemporary Sri Lanka, 'Pearl of the East' or 'Island of Tears'?*, 217–228. Sage, New Delhi.

McGilvray, D. 2001. *Tamil and Muslim Identities in the East.* Marga Monograph Series on Ethnic Reconciliation No. 24. Marga Institute, Colombo.

O'Callaghan, S. & Pantuliano, S. 2007. *Protective Action: Incorporating Civilian Protection into Humanitarian Response.* HPG Policy Brief 29. Humanitarian Policy Group (HPG), London.

Pantuliano, S. & O'Callaghan, S. 2006. *The 'Protection Crisis': A Review of Field-based Strategies for Humanitarian Protection in Darfur.* Humanitarian Policy Group (HPG) Discussion Paper. Humanitarian Policy Group, Overseas Development Institute, London.

Peebles, P. 1990. Colonisation and ethnic conflict in the Dry Zone of Sri Lanka. *The Journal of Asian Studies* 49:1, 30–55.

Philipson, L. & Thangarajah, Y. 2005. *The Politics of the North-East. Part of the Sri Lanka Strategic Conflict Assessment 2005.* Asia Foundation, Colombo.

Rajasingham-Senanayake, D. 2002. Identity on the borderline: Modernity, new ethnicities and the unmaking of multiculturalism in Sri Lanka. Silva, N. (ed.) *The Hybrid Island: Culture Crossings and the Invention of Identity in Sri Lanka*, 41–70. Zed Books, London.

Rampton, D. 2007. *Development, Humanitarianism and Spectre of Colonization in the Eastern Province.* Paper prepared for the international seminar Humanitarian Action in the 'Undeclared War' in Sri Lanka, 22 September 2007, Geneva, Switzerland. http://www.dev-zone.org/downloads/DavidRampton%5B1%5D.pdf (accessed 20 November 2007).

Schrijvers, J. 1999. Fighters, victims and survivors: Constructions of ethnicity, gender and refugeeness among Tamils in Sri Lanka. *Journal of Refugee Studies* 12:3, 307–333.

Slim, H. & Bonwick, A. 2005. *Protection. An ALNAP Guide for Humanitarian Agencies, Active Learning Network for Accountability and Performance in Humanitarian Action.* Overseas Development Institute, London.

Stirrat, J. 2006. Competitive humanitarianism: Relief and the tsunami in Sri Lanka. *Anthropology Today* 22:5, 11–15.

UN Security Council. 2007. *Report of the Secretary General on the Protection of Civilians in Armed Conflict.* S/2007/643, 28 October 2007. United Nations, New York.

UNHCR. 1998. *Report of the Representative of the Secretary-General, Mr. Francis M. Deng, Submitted Pursuant to Commission Resolution 1997/39. Addendum: Guiding Principles on Internal Displacement.* 11 February 1998. E/CN.4/1998/53/Add.2. UN Commission on Human Rights, Geneva.

UNHCR. 2000. *The State of the World's Refugees.* Oxford University Press, New York.

UNHCR. 2007. *Global Appeal 2008-2009, Sri Lanka.* http://www.unhcr.org/publ/PUBL/474ac8dc0.pdf (accessed 21 February 2008).

Uyangoda, J. 2007. *Ethnic Conflict in Sri Lanka: Changing Dynamics.* Policy Studies 32. East-West Center, Washington DC.

# Whose solution? Policy imperatives vis-à-vis internally displaced persons' perceptions of solutions to their situation in the Sri Lankan conflict

SUZANNE Y.A. TETE

The Guiding Principles on Internal Displacement proposed by Francis M. Deng in 1998 offer three durable solutions to internal displacement. States are not unwilling alliances when it comes to drawing up elaborate policies as solutions in the best interests of their displaced populations, but which also, albeit most importantly, help to preserve the fragile national security situation of which IDPs are often a product. By counter-posing such dominant government and/or policy discourses on return to the lived realities of IDPs in a dynamic war-torn context, the paper seeks to highlight the 'subaltern' narratives of conflict-displaced IDPs in Vavuniya, in northern Sri Lanka, with regard to the solutions they consider viable in the face of ongoing conflict. Policy narratives and the discursive practices they produce, as well as an eclectic consideration of various concepts, should be useful in understanding not only the policy practices but also the extent to which the displaced are (or are not) part of the solutions that claim to address their displacement. Accordingly, a qualitative method of enquiry was judged to be a relevant approach to answering the question of whether the voices of internally displaced people (IDPs) feature in the discourses claiming to provide a solution to their displacement. The method was applied in an attempt to bring to the fore the perceptions and meanings which IDPs in Vavuniya assign to policies implemented on their behalf.

*Suzanne Y.A. Tete, Department of Geography, Norwegian University of Science and Technology*

## Introduction

Taking as a starting point an excerpt from an interview with a Tamil IDP in 2007 (Box 1), I wish to examine, from the perspective of Internally Displaced Persons (IDPs) in the north-eastern district of Vavuniya in Sri Lanka, some practices based on 'the right to return' and the relevance (or otherwise) of such a 'right' in IDPs' situations, and also to better inform the ongoing discourse on this 'right'. Highlighting the pertinence of adding IDPs' voice to the policies implemented on their behalf, as attempted in this paper, is especially important for several reasons, a few of which will be discussed in the following.

According to the Internal Displacement Monitoring Centre (IDMC 2008), in 2008 the global IDP population stood at an estimated 26 million in at least 52 countries, constituting a 6% increase over the figures for 2007 and marking the highest figure since the early 1990s. Of the IDPs, an estimated 11.3 million in at least 13 countries were without any significant humanitarian assistance from their governments; whilst 21 countries had governments or occupation forces directly or indirectly involved in deliberately displacing people. Further, an estimated 9.3 million in at least 10 countries are faced either with governmental indifference or with hostility toward their protection needs. Yet IDPs, by virtue of not having crossed an international border, do not have special legal status comparable to that of other forced migrant categories, such as refugees, and are actually considered to still be under the protection of their own state – the same state that in some cases has been unwilling and/or unable to care for them and might even be an important cause of their migration flows.

To complicate matters further, IDPs are the result of overlapping causes and consequences ranging from violent conflicts to gross violations of human rights, and to natural and human-made disasters. Some are returning refugees who sometimes return only to be re-displaced internally due to recurring and/or long-standing conflicts, as is the case in Sri Lanka. Thus, most IDPs are not only trapped in protracted displacement, but are often destitute, marginalised and likely to have their fundamental rights and freedoms compromised, and their economic and political opportunities destroyed. Such adverse effects spread to the communities they leave behind as well as the already strained host communities in which they seek refuge. As succinctly summarised by Brookings-Bern (2007, 7), 'it is now well-recognised that to be internally displaced is to be exposed to a range of particular risks and vulnerabilities, even if it does not create a legal status. Bringing an end to this precarious plight is critically important'.

In short, if there is a group whose plight requires urgent attention and who should be a humanitarian priority, IDPs more than qualify for such interventions. Despite best intentions and efforts to assist, however, IDPs often have to be self-reliant and also face obstacles accessing essential services and securing livelihoods in increasingly hostile environments, as is the case in the escalating conflict situation in Sri Lanka (Kälin 2000; IDMC 2008, 7–8; UNHCR 1998). Mechanisms to ensure IDPs protection, provide some assistance and search for viable solutions to their problems are thus vital if any claims can be made to promoting the development of sustainable peace. Furthermore, IDPs clearly should be important contributors to such mechanisms, rather than the exception as often turns out to

*Box 1.* A Tamil IDP's voice on 'durable solutions' in the face of ongoing conflict in Vavuniya

S. Do you have any short-term plans?
A. Nil.
S. Do you have any long-term plans, for the next 5–10 years? What are these?
A. Nothing. We will have to go to our own place, *if peace* should come to both parties in Sri Lanka.
S. Where do you see yourself in the next few years?
A. Unable to say – in Sri Lanka.
S. Do you think about the future? What do you see?
A. In the present prevailing situation, [we are] unable to think. We see an empty situation.
S. What goals have you set for yourself (if any) and what steps are you taking towards reaching them?
A. We want a peace situation in our places. We are willing, both parties (Tamil and Singhalese) to have face-to-face talks held in friendship, compassion and kindness.
S. What resources by way of assistance could help you reach these goals?
A. Nobody can do anything without peace.

*Source:* Life history account from field interview (author's italics)

be the case in practice. That such a need is widely acknowledged is reflected in a document to identify the rights of IDPs, which has been drawn up by various experts from different parts of the world. The document Guiding Principles on Internal Displacement, commonly known as the Guiding Principles or GPs (UNHCR 1998), claims to approach displacement from the perspective of the needs of IDPs (Kälin 2000).[1] In fact, in the section of the document dealing with solutions (see Principle 28), a key element appears to be voluntariness on the part of IDPs in availing themselves of any of the solutions proposed.

## *The Guiding Principles on Internal Displacement*

Drawn from international human rights law, humanitarian law and analogous refugee law, the Guiding Principles (GPs) essentially aim to consolidate the legal standards relevant to IDPs into one framework. In addition to this restatement of existing legal norms, they aim to 'address grey areas and gaps identified in the law' (Kälin 2000, v). The GPs spell out the rights and guarantees pertaining to IDPs in all phases of displacement, an important component being solutions to displacement. To this end, the GPs put forward three solutions to internal displacement, namely: return to the place of origin, resettlement in the form of local integration in the areas in which IDPs initially took refuge, and resettlement in the form of settlement in another part of the country) (see Principals 28 and 29) (UNHCR 1998). Unlike refugee law, however, they are applicable to people who are displaced within the borders of their state of origin, rather than outside (Kälin 2001; IRIN 2002; Mooney 2003).

Although not legally binding, the GPs are comprehensive and increasingly constitute an international model to adhere to when dealing with IDPs. Thus, states are actually encouraged, if not expected, to translate the GPs into national strategies or action plans and to be guided by them as they seek to address the needs of their displaced populations.[2] This is especially important in cases when states which are unable or unwilling to care for their displaced have to rely on assistance from International Humanitarian Agencies (IHAs) and other International Non-Governmental Organizations (INGOs). Furthermore, states are not unwilling alliances when it comes to drawing up elaborate policies and action plans that are presented as solutions in the best interests of their displaced populations, which also, albeit most importantly, help to preserve the fragile national security situation of which IDPs are often a product. However, in states' bid to exercise sovereignty and be seen to be protecting its citizens, do the voice of IDPs feature at all within the discourses that claim to provide solutions to their problem? A brief consideration of the official view regarding the end of displacement or what the attainment of a durable solution is supposed to ensure should prove insightful, especially when examining such views in the light of actual policy practices, as will be discussed later in this paper (see section on Policy narratives and return policies in Sri Lanka).

## *End of displacement? – official discourses*

As Kälin noted in his address to the United Nations General Assembly in 2007: 'displacement ends when one of these durable solutions occurs *and IDPs no longer have needs specifically related to their displacement*' (2007, 2; my italics). According to Kälin (2007, 3), the end of displacement is also considered to be a 'process through which the need for specialized assistance and protection diminishes progressively'. In cases where IDPs have taken advantage of any of these solutions, attention is to be given not only to the process through which the so-called durable solution is achieved but also the condition of the displaced. This 'process' would then mean that IDPs make an informed choice about the particular solution preferred without coercion, and are full participants in the planning and management of such a decision. The onus would then fall on the government to take appropriate measures to establish conditions and provide the means for such durable solutions. It would also have to grant or facilitate safe, unimpeded and timely access of humanitarian organizations and other actors assisting IDPs in this regard. Kälin argues that through such a process a durable solution is finally achieved when former IDPs can fully enjoy their human rights through certain conditions, which he lists as:

> [N]on-discrimination and equal access to legal and physical protection; livelihoods and means of survival; income generation; public services; including education, health services and pensions; political rights; the right to documentation; freedom of movement, and redress for abuses including property restitution or just compensation. (Kälin 2007, 3)

This is undoubtedly an ideal checklist, but whether it can ever be implemented is another issue altogether. In addition to considering the processes and conditions, there has been a call for an integrated approach that takes into account cause-, solution- and needs-based criteria in determining when internal displacement ends (Mooney 2003; Brookings-Bern 2007; Ferris 2007). There can be little doubt that the Guiding Principles and accompanying policy language as to the end of displacement are as extensive and ideal as possible. Yet the extent to which they reflect, albeit address, the realities on the ground must surely go beyond mere semantics. I have cited and quoted Kälin's views extensively here because they reflect to a large extent the core of the issues I wish to highlight with regards to exactly what the 'right to return' is purported to be in protracted IDP situations. In view of the elaborate articulation of the 'the right to return', not only on paper but also a certain fixation with its application and significance in policy practices as opposed to the realities that often exist on the ground, I will next briefly explore the discourse on the 'the right to return' and its translation into policy.

## *The right to return and policy practices*

The 'right to return' is a key human right principle enshrined in the Universal Declaration of Human Rights (UDHR), as stipulated in Article 13. Whilst not legally binding, this same right has been codified in Article 12:4 of the International Covenant on Civil and Political Rights (ICCPR) which gives legal force to many of the rights proclaimed in the UDHR (UN 1948; OHCHR 1976; Amnesty International 2001; Agterhuis 2005). In a broader sense, the 'right to return' is part of the freedom of movement which can be internal or within a country and external or between states. This makes it applicable to both refugees as well as IDP situations, as individuals should be able to invoke the freedom of movement either intra- or interstate. However, exercising this right, whether to enter or return, especially between states, extends beyond the individual to invoking issues of nationality, since states exercise their sovereign right to regulate the right in accord with their respective laws on and conditions of admission and readmission (Liu 2008). Speaking about a central norm informing internationally-supported voluntary repatriation processes, Bradley (2007, 3) points to 'the prevalence of return in conditions of safety and dignity'. Whilst the latter has always had an important place in refugee law, it is no less important in the GPs too (see Principle 28). This is hardly surprising since the solutions for IDPs mirror those available for refugees. Whilst the development of the 'right to return' can be traced to the post World War II era and the subsequent cold war dynamics, the reasons for the continued importance of this 'right' in displacement policy are much more complex. In this respect, what might be termed essentialist notions about place, territoriality, home, and the nation state have made a considerable contribution. The application of such concepts and the rational they provide for a policy focus on the 'right to return', will be discussed later in this paper (see section on Place, territoriality, home, and the nation state).

The main objective of this paper is to explore the extent to which government and other policy practices with respect to the right to return respond (or not) to the lived realities of IDPs and whether the latter are accorded such a right at all. Do IDPs' voices really feature within the official discourses that purport to provide solutions to their problem? What are their perceptions of the solutions being implemented on their behalf in the current context of conflict? These questions will guide the ensuing discussion. Concepts of place, home and/or territoriality, and the nation state and also the ways in which these inform policy practices such as the right to return, as well as policy narratives and the discursive practices they produce, will be conceptually useful in understanding the extent to which the displaced are a part of the solutions that are implemented on their behalf as a means of addressing their displacement. A qualitative method of enquiry is adjudged more relevant in this exercise as it seeks to highlight the perceptions and meanings which displaced people assign to such policies. Hence, in the case of the study area, extensive generalisations would, as a matter of necessity, be sacrificed for detail and an in-depth understanding of IDPs voices as to the solution which the latter consider to be more relevant in addressing their actual needs and concerns under the prevailing conditions in the Northern and Eastern Provinces of Sri Lanka. While not claiming to address all IDP situations, given that each case is unique, the experience and perceptions of Sri Lankan Tamil IDPs in Vavuniya District in the Northern Province may provide important insights into the issues they consider relevant and, by extension, the views that should be incorporated if their rights are to matter. This is especially important since the underlying assumption in all of these policies is that IDPs are or should be an important part of the planning and implementation of decisions or solutions relating to their problems. To further contextualise the discussion, I will briefly consider the background to the current conflict situation and the study area, as well as my experiences in the field with respect to the methodological implications that the experiences had for my research.

## *Background to the conflict and current context in the north-east*

The 1983 anti-Tamil riots marked an important point in what has become a protracted conflict in Sri Lanka, that has a long history and has been attributed in part to British colonial policies and practices that created fissures especially between the majority Sinhalese and the minority Tamils (Uyangoda 2007). Sri Lanka, however, is a multi-ethnic, multi-linguistic and multi-religious state, consisting of 69% Buddhists, 15% Hindus, 8% Muslims, and 8% Christians. The ethnic composition as at 2002 was 74% Sinhalese, 18% Tamils, 7% Muslims, and 1% others (Burghers, Moors, Malays, aboriginal Veddas) (Lund 2003; USDS 2007; Uyangoda 2007).

In an ongoing conflict which is said to have killed more than 70,000[3] people (Uyangoda 2007), displacements are common – both external and also mainly internal. With regard to the source, nature and intensity of such displacements, Hyndman & de Alwis succinctly note the following:

> [T]he *widespread* and *repeated* displacement experienced by *hundreds of thousands* of Sri Lankans, especially Tamils, is a

> *very visceral process* of forced migration generated through *geopolitical struggles* and managed further through technologies of the state ... that include identification cards and checkpoints along major roads that connect Sri Lanka's towns and cities. (Hyndman & de Alwis 2004, 536; author's italics)

The tsunami of 26 December 2004, with its attendant devastation, mostly in the north-east of Sri Lanka, compounded the existing inequalities caused by conflict and has led to even further marginalisation of IDPs in these areas. This is the case despite the GPs' stipulation that all IDPs should be treated equally regardless of the cause of displacement and despite competent authorities' claim to honouring their obligations in this respect (see Principles 6.1a, b, and d). What is widely documented, however, are the inequities in treatment occasioned by the international attention and subsequent aid directed to the tsunami-displaced as against the conflict-displaced, especially in the worst-hit area, the north-east (Martin & Charny 2005; Sandlund 2005; Amnesty International 2006).

With respect to the present situation, it has been observed that:

> [T]he current government's pro-Sinhalese nationalist policy of massive military retaliation against the separatist Liberation Tigers of Tamil Eelam, (LTTE) ... has contributed to gross human rights abuses and increased misery especially among the Tamil civilian population in the North-Eastern [Northern and Eastern] Provinces of Sri Lanka. (Uyangoda 2007, 68)

Furthermore, the effects are said to be especially pronounced 'in the LTTE-controlled areas mainly in Mullaithivu and Killinochi, and also parts of Jaffna, Mannar, *Vavuniya*, Trincomalee, Batticaloa and Ampara districts' (Uyangoda 2007, 68; my italics). This is the reality for the thousands of IDPs located throughout the north-eastern parts of Sri Lanka and the context in which my fieldwork was undertaken. As noted by the International Committee of the Red Cross (ICRC 2008),

> [T]he ceasefire agreement of early 2002 officially ended in mid-January 2008, though fighting between the LTTE and the Sri Lankan Army (SLA) had resumed in mid-2006 along the east coast and on the edges of the Vanni. The number of casualties among civilians has been sharply increasing. There has been a dramatic increase in the number of alleged International Humanitarian Law (IHL) and human rights violations and of mass population displacements in the North and East.

Pointing to the worsening situation occasioned by the renewed fighting, the United Nations High Commissioner for Refugees has indicated that the total number of IDP movements after 7 April 2006 was 187,863, of which 10,657 IDPs that had originated from the districts of Vavuniya, Jaffna, Trincomalee and Mannar were displaced to the Vavuniya District[4] (see also IDMC 2007; IRIN 2008). This indicates, not the fixity, but what Malkki (1997, 86) describes as 'the transitory, deterritorialised, unfixed and processual nature' of Vavuniya as an IDP generating and receiving transit point, to be studied as such. The figures are as comprehensive as can be expected, yet behind these statistics there are individuals and families suffering the brunt of the senseless attacks brought to bear not only on combatants but also on civilians who are used as human shields and objects of retaliatory or reprisal attacks by both the LTTE (Liberation Tigers of Tamil Eelam) and the Government of Sri Lanka (GoSL). This is in sharp contrast to the relative but fragile calm that existed from the signing of the much-acclaimed Memorandum of Understanding (MOU) in 2002 until the escalation of violence in June 2006 (Uyangoda 2007). In fact, this was a period during which many IDPs felt encouraged to explore the possibility of returning to their places of origin, and many did so. According to Newman (2003, 5), 'by the spring of 2002, an estimated 260,000 displaced Sri Lankans had returned to their homes with rates of return continuing at a similar pace as at 2003'. Long-term development and rehabilitation projects for returning IDPs (UNHCR 2003) as the policy direction in terms of finding solutions for the displaced may have been appropriate then, but may not be viable in the current environment of renewed fighting, which is marked by military and counter-military attacks, constant shelling, and prevailing conditions of generalised insecurity as borders of cleared and uncleared areas[5] continue to be redefined through active warfare by the government and LTTE alike.

### *Study area*

Vavuniya in the Northern Province of Sri Lanka is affected by conflict, in common with many districts in the north and east. It has four District Secretariat (DS) Divisions, one of which is controlled by the LTTE. Vavuniya is an interesting site for studying IDPs due to its location close to the conflict zone; it is a border district with a Forward Defence Line (FDL) between the government forces and the LTTE, a position which makes it both an IDP generating and receiving district. According to Van Hear & Rajasingham-Senanayake (2006, 47), '[As] in Mannar, Trincomalee and Mullaitivu Districts, IDPs come and go depending on the security situation, military operations and the conditions of shelter'. Field visits were made to Poonthoddam Welfare Centre (WC), one of several transit or temporary centres found throughout the north-east, which are camps where IDPs have been displaced, often for years. Visits were also made to Kalnaddinakulam, Thaddankulam, Kandasamy Nagar, Manik Farm, and Kalmadu Relocation Villages (RVs) in various DS Divisions of Vavuniya District (Fig. 1).

## Methodological and theoretical considerations

### *Researching under fire: impressions from the field and methods of data collection*

The field was characterised by the following: military camps; soldiers milling around; guns pointed at potential targets – imagined or real; gunshot and its characteristic smell; constant shelling; endless checkpoints; stories of disappearances, abductions, suicide bomb explosions, reactionary blank-range shootings; the imminent possibility of being

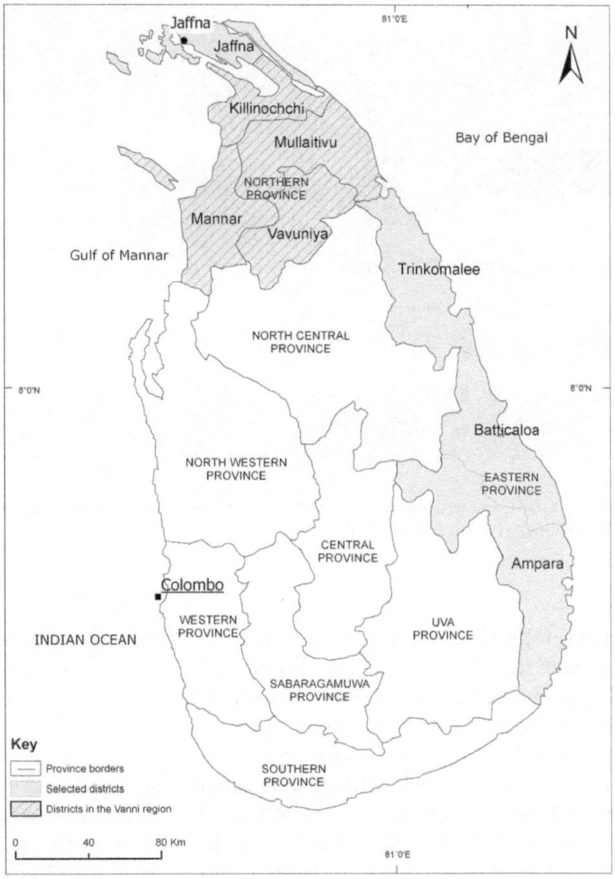


*Fig. 1.* Sri Lanka: location of study area (Vanni region)

caught under fire; the need for extra caution in the most basic undertakings; and civilians on edge but maintaining an admirable calm on the surface, striving to go about their day-to-day activities as best as possible. However, such experiences do not even begin to express the nerve-racking feelings of fear, insecurity and uncertainty that can be and is occasioned by people's efforts to go about their daily lives in the heat of war (Figs. 2 and 3).

*Fig. 2.* GOSL Soldiers on patrol (Photo: S. Tete, 2007)

*Fig. 3.* One of numerous military points (Photo: S. Tete, 2007)

The above gives a brief insight into the context in which I carried out a reconnaissance field trip in Vavuniya District in the Northern Province of Sri Lanka in 2007. The specific context is the period from the LTTE attack on the Anuradhapura military base to the assassination of top LTTE Cadres, including Tamil Selvan, the political head and chief negotiator of the LTTE. Vavuniya is a border town and a launch site for missiles and counter-missiles to and from the LTTE-controlled Vanni region. The study context is to be considered within the general context of protracted and escalating war between the Sinhalese-dominated government forces and the Tamil-dominated LTTE. Conducting fieldwork in such an IDP setting, with ever changing dynamics and coming under fire, has its own set of challenges, not least of which are methodological and ethical issues,[6] especially where access is restricted and one has to work within a framework of insecurity. As Newman (2003, 8) noted of her research in the northern provinces of Sri Lanka in 2002–2003, during the ceasefire, 'gathering and documenting narratives through interviews and conversations was a particularly challenging process given the tumultuous social and political timing of my research, and it clearly had a pronounced effect on both the content and form of many narratives'. In such an environment, where researchers are not the most welcome of persons, the importance of gatekeepers in gaining access to displaced populations cannot be overemphasised. A Norwegian–Swedish based development agency, FORUT-Sri Lanka, made it possible for me to conduct research, whilst FORUT-Vavuniya facilitated access to the people, the organisations and the field in general. I was also fortunate to have been able to tap into the vast experience of our project leader, Ragnhild Lund, due to her many years of research in Sri Lanka. Whilst enabling me access, my association and introduction into the field by an NGO (FORUT) was accompanied by its own challenges, not least of which were biases in terms of who I interviewed. Most of the research participants were direct or indirect beneficiaries of FORUT programmes since FORUT was a local implementing partner in the construction of Relocation Villages (RVs) and/or engaged in various programmes for the IDPs. Research participants' expectations and the perceived role they assigned to me due to my associations and the fact that I was regarded as an 'outsider', as a female African researcher, may have engendered unequal power relations and affected the responses given.[7]

However, I made sure the research participants understood who I was and the purpose for which I was there, which, although not providing any material advantages, was to try and highlight their views as a means of bringing their voices to the fore. I also requested their consent and voluntary participation, making a point of explaining the possible uses that I might have for the interviews and pictures, and seeking their permission in this regard. Their responses, like the manner in which they were coping with the stressful situation, were gracious. I also assured them of confidentiality with respect to any information they gave. However, this may have been comprised by my reliance on a local interpreter.[8] Even though he reassured me that it afforded him a rare opportunity to be in the field, listening to the IDPs, and evaluating the effectiveness of the NGO's programmes in terms of whether they were addressing the concerns which the beneficiaries had – I had to trust his word when it came to maintaining the confidentiality of the information gathered. Having to filter the information through an interpreter, however skilled he or she is, presents it own set of challenges, not least of which is the possible 'loss of nuances and personal reflections by the research participants through the act of translating. As further observed by Brun (2003, 52), translation 'often introduces different concepts and perceptions ... [and] displaces and may even do violence to local knowledge through language' (see also Alarcon 1994; Hyndman 2000).

I also had to respect the constraints placed on me by the security situation. This called for strict adherence to time schedules, as we had to be back at the hotel whilst there was still daylight, in order not to compromise our safety or that of all the staff with whom we travelled to and from the field. In addition, whilst travelling in an NGO vehicle offered some protection, nothing could assuage the fear of being caught in the middle of action, of moving through a mass of gun-pointing soldiers whilst listening to stories of disappearances, possible bomb explosions and reprisal shootings in which we could be caught at anytime, hearing thunderous peels from nearby shelling which shook us to the core, or of finding myself in a vehicle with Tamil staff but without the NGO logo and being stopped and checked by Singhalese government soldiers, or of inhaling the smell of cordite gunpowder from gunfire while in the refuge of my hotel room (Fig. 4). Such experiences were a few of those specific to the context of my research.

Incidental sampling and the snowball method of sampling, coupled with the flexibility that is characteristic of qualitative research and particularly of the life history method (Thompson 1981; Skonhoft 1988; Repstad 1993; Valentine 2001; Creswell 2003), all proved to be very important elements in accessing research participants. These approaches also afforded me the opportunity of taking into account the situation and arising issues as the research unfolded and to readjust my research problems and objectives to reflect the realities in the field accordingly (Limb &

*Fig. 4.* GOSL soldiers positioned on street behind the author's hotel window in Vavuniya (Photo: S. Tete, 2007)

Dwyer 2001). I collected five life history accounts from three women and two men, all Tamil IDPs, and these were supplemented by group discussions with members of various community-based organisations (CBOs). The members represented a range of 'return' experiences as some lived in RVs, some were self-settled, and others had lived and/or were living in Welfare Centres (WCs). These interviews were supplemented by key informant interviews with the UNHCR (United Nations High Commissioner for Refugees) and Norwegian Research Council (NRC), observations, informal conversations, pictures, my participation in debriefing and report-back meetings to FORUT, a livelihoods workshop with various implementing partner CBOs of FORUT, and also a desk search for and use of other secondary literature sources.

In view of the exploratory nature of the fieldwork, coupled with the constraints and formidable challenges presented by the worsening security situation in which it was carried out, adaptation, flexibility and what Jacobsen & Landau (2003), 196) disapprovingly refer to as 'convenience' sampling, were necessary, rather than an escape mechanism if respect, caution and safety were to be prioritised over a foolhardy all or nothing stance which might have yielded more interviews no matter what the cost. I am wary, though, of making any generalisations from the interviews as I recognise the limiting impact of the techniques on the form and content of the information that could have been collected under such circumstances. Yet as already indicated, and as Malkki rightly notes of working in such transitory research contexts, perhaps objectivity, representativeness and replicability in the methods applied should not be such a central concern as the need to be honest about 'the inevitable doubts and difficulties encountered in any act of discovery' (Malkki 1997, 99). This is not to say that anything goes as to the techniques to be employed in collecting data in such a context, but rather that trying to be an attentive listener and recognising the 'situatedness' of one's intellectual work (Haraway 1991) and 'affirming one's own connection to the ideas, processes, and people one is studying'

(Malkki 1997, 96), as I have sought to demonstrate in the discussion above, are more important in this kind of study.

Thus, the stories told by individuals, the group discussions, the key informant interviews, the informal conversations, and the direct observations all provided useful insights and details as to the concerns of the IDPs about their right to return and, hopefully, the solutions they considered viable in addressing these. If I had the opportunity to return to the study – a possibility that has been rendered impractical by the worsening conflict – I would collect more stories, stay longer, participate more as an observer, and hope that more of the IDPs voices would be projected and heard regarding the solutions purported to be for their benefit.

In the following, I will present the eclectic theoretical underpinnings of the discussion, highlighting the role of policy narratives in official articulations and IDPs' perception of return policies in Sri Lanka.

## *Place, territoriality, home, and the nation state*

As has been noted by Robinson (2002, 3), understandings about place are vital to studies of displacement and conversely, 'the phenomenon of displacement does not diminish the significance of territories or places'. Indeed, implicit in the term displacement is 'place' and the idea that people are, as the prefix 'dis-' suggests, 'not in place'. Place is thus relevant in discussions about forced migrants. It is the very nature of place that confers the label and the practices that govern forced migrants as ones who are 'uprooted' and in need of international or national protection. It is generally accepted that a person leaving his or her country may do so for various reasons, whether proactive or reactive (Richmond 1988; 1994; Turton 2003).[9] Yet, there is wide acclaim and an almost taken for granted assumption that return is usually 'motivated by a desire to return "home", to the place where one belongs or to one's roots' (Agterhuis 2005, 2). It has even been claimed that 'there is no greater sorrow on earth than the loss of one's "native" land [homeland]' (Agterhuis 2005, 4) (see for instance Nseroko 1981, and *Media* v. 650–651 (Johnston 2008)). As has been argued by Sepulveda (1995, 85), such territorialised notions of home are premised on the existence of 'a singular, immutable bond between people and a particular space', and further that such notions assume that 'people possess static, sedentary identities', and that 'flight and exile imply a rupture in the symbolic representation of this bond which links people to land' (see also Newman 2003, 13). Various return policies would then be justified as being the sole means through which to heal such 'severed bonds'. In fact, these assumptions not only guide much of the theorising in migration studies, but also the policies and practices implemented on behalf of the displaced.

Such rooted notions about place in a highly territorialised world are no doubt given salience by the modern political organisation which is built on the idea of a nation state, of which territorial boundaries are an important defining feature (Turton 2002). It is this highly territorialised nature of the nation state, central to which are the ideas of home, land and security (Walters 2004), which defines and reinforces much of the practices for and against forced migrants. As Turton (2002, 20) points out, 'forced displacement is both a threat to, and a product of the international system of nation states which seeks to reproduce itself as a viable and morally legitimate political community'. Robinson (2002, 3) thus argues that 'the nation state usually does not only cause displacement but also assumes responsibility for attempting to reduce the undesirable effects of its consequences', thus preserving the image that 'the national order of things' (Malkki 1992) is 'the normal order of things' (Turton 2002, 20).

In so far as a nation state system exists, a highly territorialised sense of place and an essentialist notion of home remain the reality in policy and in practice. This lends weight to Kibreab's argument that we live 'in a world where rights such as equal treatment, access to sources of livelihoods, social services, rights of freedom of movement and residence, etc. are apportioned on the basis of territorially anchored identities'. Thus 'the identity people gain from their association with a particular place is an indispensable instrument to a socially and economically fulfilling life'. Undoubtedly, as Kibreab (1999, 2–4) states, 'state territories – even within a state – are not and have never been there for the taking'. This, in effect, is what Walters (2004, 241) describes as 'domopolitics' or an obsession with governing a nation as a home, making it a place which belongs to 'us' and not 'them', and furnishing the nation state with its rational for excluding some whilst including others (for a detailed definition see Turton 2002, 22). That this is the case is well demonstrated in the context of an ethno-political conflict such as that in Sri Lanka, where lines are sharply drawn as to who belongs 'home' and who does not. Here, home could mean belonging to the dominant religio-nationalist group, having political and/or ethnic affiliation, or simply geographical location. These and other such factors could and do determine not only the type of solution promoted for various categories of IDPs, but also the form of assistance provided. For instance, as Van Hear & Rajasingham-Senanayake note about the range of experiences of displacement by various categories of IDPs in the north-east, 'the living conditions and life chances of displaced people ... vary greatly according to the *geographical area* in which they found themselves ... according to the reach of the humanitarian regime, and according to the duration of displacement' (Van Hear & Rajasingham-Senanayake 2006, 59–60; my italics). Although the concepts informing policy implementation for IDPs are much more complex than those discussed above, understandings about place, territoriality, home, and the nation state play an important role in the emphasis on the 'right to return' and its variants as an ultimate solution to displacement. Further, this conceptual framework remains the major discourse informing scholars of forced migration and the international humanitarian community (IHC). However, as Newman (2003) argues, it has yet to comprehend the complexity of the experiences involved in return and the attendant processes. I will further contextualise and exemplify the somewhat abstract theoretical discussion above by considering how competing policy narratives have guided return policies in Sri Lanka and are perceived by the IDPs.

## *Policy narratives and return policies in Sri Lanka*

According to Blaikie et al. (2007, 93),

> [P]olicy narratives are stories told by different protagonists . . . as a way of making sense of an uncertain, complex and contested world or more strategically, as a means of persuading others . . . these are not just tales or inventions for others' amusement but persuasive constructions which frame issues and problems in certain ways to focus on some issues and to exclude others. Thus some voices may not be heard and are unable to join the negotiating table or their knowledge may not be deemed by more powerful actors to be worthwhile or legitimate.

This may yield policy processes and outcomes which are as elaborate on paper but which alienate those whom they are intended to benefit.

In Sri Lanka, and as stipulated in the Guiding Principles, the Government is the competent authority (though not the only party) primarily responsible for enacting and implementing policies with respect to IDPs. Its policy with respect to return has been resettlement (the return of IDPs to their original homes) or relocation, i.e. settlement of IDPs on state or private land (WFP 2000; Van Hear & Rajasingham-Senanayake 2006). Whilst returning home is the option most favoured by both the government and the IHC, the former has tended to stress return even when other options may be more appropriate. Today, it is only the heat of unabated conflict that seems to have stalled its implementation, though it is by no means an abandoned project. Depending on the dynamics of the war and other factors, relocation to other areas is undertaken to a limited or greater extent, but with restrictions; whereas local integration is usually not permitted, the displaced often prefer to be allowed to stay where they have been displaced, especially in Vavuniya town.

In its preference for return, the GoSL might be trying to respect IDPs' own expressed wish to return (see the section on Return) or striving to fulfil its primary obligation of protecting IDPs and also endeavouring to act in line with a policy narrative that favours return. Yet whilst IDPs have expressed a desire to be able to return eventually and/or to be relocated, they have also been clear on their need to do so in a context of peace rather than that of the raging war in which these policies are currently implemented (see the section on Insights from the field). Hence, the possibility that political and ethnic considerations are important factors behind these decisions should not to be ignored. This is especially highlighted by the government policy stipulating that IDPs cannot be relocated on state or private land in any district other than that of their origin. Affirming this, the Centre for Policy Alternatives CPA (2001, Summary) states: 'the Government policy does not allow the displaced in its relocation programmes to settle outside their Districts of origin, which in most cases are too unsafe to return to' (see also Jayasinghe 2005; Banerjee 2006; Van Hear & Rajasingham-Senanayake 2006). Apparently, there is an inherent contradiction in the geopolitical agenda which impacts negatively on the narrative and practices relating to 'the right to return'. As IDMC (2006) and Amnesty International (2006) have observed, this geopolitically powerful narrative could be more concerned with not upsetting the ethnic status quo, so that the make-up of the electorate wins over any humanitarian considerations. The IDMC pithily notes, 'relocation policy will *compel* landless IDPs to either remain in camps or return to an area against their will' (2006, 161; my italics).

Relocation, however, was the policy practice observed in the course of fieldwork, despite the formidable challenges presented both to the displaced and the plethora of I/NGOs struggling to maintain a presence in a not-so-welcoming context. These range from the remoteness of RVs from the nearest main town, to having to cross numerous military camps of tense gun-touting soldiers, enduring constant checks which sometimes last for hours if not days, easy abductions by all groups, and also recruitment of children by paramilitary groups, disappearances, generalised insecurity and fear with their attendant psychosocial problems, and limited economic options partly due to restricted access. Added to this is the poor infrastructure of some semi-permanent houses provided and the difficulties occasioned by housing located in waterlogged areas, to mention a few. Considered in the light of the GPs and the expectation of states to abide by them, can it be claimed that pushing ahead with the current policy practices in this situation of ongoing conflict and its attendant insecurities is in the interests of the displaced? Further, are they ensuring the voluntariness, safety and dignity explicitly required for the attainment of any sustainable solution to people's displacement?

*Policy narratives and the 'subaltern' voice.* As is succinctly noted by Blaikie et al. (2007, 93–94),

> [A] focus on narrative in the policy process is necessary but insufficient in addressing the reflexive relationship between the power of the author of the narrative and the narrative itself. . . . a focus on the policy narratives alone will tend to privilege documents, printed matter and the survival of the formal and the well-financed over the spoken narrative by those who have poor access to a printing press, radio or politically powerful actors at different levels.

This points yet again to what the GPs admit to be worthy of special attention and effort: 'the full participation of IDPs in the planning and management of their return or resettlement and reintegration' (Principle 28.2). Thus, whilst recognising the importance of the structures and institutions involved in drawing up and shaping policy with respect to the 'subaltern' actors[10] (Spivak 1988; Bhabha 1996), I wish to highlight the narratives of those without whom one cannot speak of a solution, but whose voices may not be heard as loudly as others (Blaikie et al. 2007). In the ensuing discussion, I present excerpts of narrative or life history (LH) accounts and insights gleaned from other conversations with some IDPs in Vavuniya as to their lived experiences and their perceptions of the main policies being implemented as 'solutions' in the current context of conflict.

## *Insights from the field*

The proof of the effectiveness of any policy, especially if it is to be called a durable solution, can and should be assessed at the local level. Yet, that the views of IDPs should count in such decisions seems much more elaborate in theory than is often the case in practice (see for example Principles 7b, c, and d). Accordingly, the concerns expressed by the IDPs presented below should aid in understanding the meanings they assign to such policy practices and hopefully highlight what they are whispering, if not shouting, as to the way forward.

## *Life in a welfare centre*

> Initially the police post was here. Now it is no longer here. Now anybody come and roam around this place ... during the night, we cannot identify. There is no security for us here. Sometimes the security forces come and round up this area. Electricity is very important but it is no longer here, it is also cut off and it affects the education of the children. When it comes to advanced level students the situation is much worse because they cannot go for other means to get light; it is expensive. Alcohol consumption is another problem. There is also teasing when we (girls) pass. (LH account of an IDP youth on life in Ponthoddam WC)

As indicated in the account presented above, life in the WCs is increasingly traumatic since whatever little protection there was has since been withdrawn. At the time of our visit, this concern had still not been addressed. That this disturbing situation still pertains is indicated by a recent IASC (2008a) situation report on Mannar and Vavuniya stating that the, 'UNHCR reopened the issue of having police presence at the WCs of Poonthoddam and Sithamparapuram in Vavuniya'. With the absence of light and the police, it has become an ideal recruiting ground for various elements.

Educational opportunities are severely restricted and where people have been able to attend classes, the environment is hardly conducive to any serious studying, given that IDP students have to compete with other children who are not in the same situation (Jayasundara 2007). Idleness and attendant problems of alcohol abuse, domestic violence, and child and teenage marriages are increasingly common. This makes it a further fomenting ground for political fractioning and ideal for recruiting youths and other frustrated residents. However, some prefer to be at WCs for the ease of access they afford them to the nearest town, to which their livelihood activities are connected:

> Actually we don't like to go to faraway places for relocation. That doesn't mean we like to stay in the WCs. But even though it is a WC, it is in the town, so our livelihood activities are always connected to the town area. So if we go to faraway places we loose that opportunity. So in that sense, we have a negative view. (Group discussion with a women's CBO)

Generally, however, many in the WCs expressed the hope of being considered for relocation to one of the villages, even if this presented them with dilemmas due to their being in 'faraway' location (as opposed to the proximity of the WCs to town). This was especially because the WCs were increasingly becoming unsafe. Thus, wanting to remain there appears to be a matter not of choice but, of being between a rock and a hard place.

## *Relocation*

> Nellikulam camp was actually so crowded it was not a healthy condition to live in. It was a big shed partitioned by UNHCR tarpaulins/canvas. Inside it was very hot. There was no room, only one space portioned by canvas (between family units) and a common toilet. Taking a bath was also difficult. There was no land for our personal activity. For income we had to go out and work and our parents had to get money for our essential items. But when you compare Thaddankulam RV, it is different because at least there is some land for our thing and we can have some backyard poultry/home garden and activities like that for our food security and small income. (LH of an IDP youth on life in the WC as opposed to the RV)

Whilst many were happy to have been assigned land in the relocation villages as a respite from the terrible conditions they had endured in their years at the WCs, the conflict situation today has compounded their vulnerabilities and uncertainties (Figs. 5a and 5b). This has contributed to their feeling of being trapped. Initially, they were grateful to own pieces of land, even if only a half or quarter of an acre (0.1–0.2 ha), for the opportunity it afforded them to cultivate

*Figs. 5a* and *5b* Partitioned sheds in Poonthoddam Welfare Centre (Photo: S. Tete, 2007)

crops and to enjoy some privacy to an extent, in contrast to the limited space and restrictions characteristic of their stay in the WCs. Following the breakdown of the peace accord however, there has been heightened feelings of insecurity among IDPs, especially as they live surrounded by the army and other non-state actors. This imposes severe restrictions on their movement and livelihood opportunities. Venturing out to work is rather difficult since military checks are the order of the day. In the likely events of bomb explosions, mines detonating and the abduction of young people, even a simple ride to school presents incredible challenges. This explains the sense of foreboding on the part of IDP parents as to what the future holds for their younger children. Older women have had to assume the role of caregivers in the absence of their daughters, some of whom have migrated to India as refugees or to the Middle-East to work. This compounds existing vulnerabilities even though ingenuity and a desire to engage characterises the way in which many are striving to cope under such daunting circumstances.

Advantages and disadvantages of life in a Welfare Centre (WC) compared to a Relocation Village with respect to the dilemmas presented by the situation today are summarised in Table 1. The residents of one RV considered life in the WCs as the lesser of two evils. Rather than an indication of a real interest in returning to these centres, a more accurate reading is offered by the adage 'the devil you know is better than the angel you do not know'. Evidently, the formidable challenges presented by living in a remote RV as compared to living in WCs accounted for such views. The WCs were at least centrally located and relatively safe, even if the residents might have been unaware of the unsafe security situation in these WCs too.

Poor infrastructural development in certain RVs resulted in the uneven situation of some residents of the same village being housed in waterlogged areas and hardly able to even access their toilet facilities. In addition, there was inequality in the construction of semi-permanent houses and its attendant problems such as leakages during the rainy season, in contrast to the provision of permanent and better built houses for some IDPs but not others, and also discriminatory selection of beneficiaries (Figs. 6 and 7).

*Table 1.* Advantages and disadvantages of life in a Welfare Centre (WC) compared to a Relocation Village (RV)

| | Positive | Negative |
|---|---|---|
| Life in a Welfare Centre (WC) | | |
| | Dry rations | No space |
| | Police and security | No privacy |
| | Electricity | Not good for children |
| Life in a Relocation Village (RV) | | |
| | Land | No food rations |
| | Privacy | No protection |
| | | No electricity |
| | | Fear of being outside |
| | | Fear inside, particularly during night-time |
| | | Loss of hope for a better future |

*Source:* Ragnhild Lund (Project Leader)

*Fig. 6.* Residents trapped by water in a Relocation Village (Photo: R. Lund, 2007)

All of the aforementioned practices also raise important questions about the interests at stake, as to whether efforts are – as stipulated by the GPs and the dominant policy discourse – really about finding solutions that place the dignity and safety of IDPs above national interests or whether the reverse is rather the case. Whilst the government's efforts at providing land for relocation of some displaced people are commendable, the question remains as to the practicality of such locations, which have become traps rather than the solution they were meant to provide. The pragmatism in pushing ahead with such exercises in the present-day context of ongoing conflict and widespread insecurity is thus highly contested and leaves much to be desired.

## *Return and perceptions of home – 'Home is "own soil"'*

We want to say look at another angle . . . even though now we say Jaffna and some other places are our home, what I mean is, although the period [we have lived here] is short we were living very happily. That is a very happy life and also we were living with

*Fig. 7.* Waterlogged area with semi-permanent houses in a Relocation Village (Photo: R. Lund, 2007)

our own relations but here the situation is not so. (Group discussion with members of an IDP women's CBO displaced mainly from the Vanni and living in various areas of Vavuniya)

The above views, expressed in response to where the IDPs considered home to be, resonated in other narrative accounts as to why return was envisaged, though certainly not desired under the current circumstances. Ultimately, home for many was the place where they were born and/or were settled before they were chased out. As expressed in the following, this was either by direct conflict with shelling and the real threat this posed to their lives, or because their villages were taken over as high security zones (HSZs) by the army and thus rendered uninhabitable:

> We left Jaffna in 1995 when it was suddenly announced that the army was at Jaffna Penninsula ... We were displaced to Kilinnochi and Mullaitivu but had to leave each time because the army was coming, shelling commenced and it wasn't safe ... It was a dangerous journey and a very difficult crossing into Vavuniya after we left the LTTE-controlled area. (LH account of an IDP youth on her displacement experience)

> I am from Omanthai in Vavuniya District. I was displaced from there due to shelling and heavy fighting between LTTE and government forces; there was intense gunfire when fighting erupted at night so we couldn't stay anymore and had to leave. My father had assets in Omanthai and was rearing some livestock, so due to the movements he became frustrated as he was thinking a lot about all he had lost. So he passed away three months after our going to the WC. (LH account of an IDP woman's displacement experience)

> I was born in Kandy but moved to Killonochi at about three years of age due to my mother's remarriage to a man from there. Our house was in the elephant path but we lived there until it was later declared an army high security zone, but now it is under LTTE control ... we moved to the WC in Vavuniya due to high intensity war. (LH of an IDP woman on her displacement experience)

As the situation still persists, it is one of many reasons why return is not viable. Many IDPs expressed the desire to be able to return to their place of origin for the various advantages this would offer. Whilst nostalgic feelings may and often do have their place in such articulations about home, the readiness of some IDPs to return spontaneously and at great risks to themselves calls for a pause in dismissing such perceptions as mere wishfulness or wistfulness on the part of the displaced 'to reconnoitre or take the security temperature, and if not too dangerous, to return for a while at least, with or without their families' (Clarance (2007, 233). As to a sampling of the reasons why return was envisaged, the views expressed below, though not exhaustive, are instructive:

> Home is my own place ... that means '*own soil*'. When we say own soil, we don't think too much about the soil but we talk about original place ... that is *our original ownership.* You have mentioned the period grown up with parents, but actually we're thinking in the other direction ... *the quality of life.* Because when you talk about the long period we spent in Vavuniya WC, there we were living in 10 square feet [0.9 $m^2$] sheds ... whereas although it was for a short period, we were living in a big space at home. That is why we still consider that as our home, not the period, but the quality. Yes, we feel that at another angle when we look at the restrictions. There we didn't feel any restriction, we had lot of freedom. But here ... our space is limited, that is a restriction and there are several other restrictions like that. (Group discussion with an IDP Women's Group displaced mainly from the Vanni and living in various areas of Vavuniya; my italics)

Invariably, the desire to return was linked to the loss of land, property and the means of livelihood that these supported. Also missed were the social network and family support systems, and the generally better conditions that were perceived as having prevailed in such places but that had been lost, as well as the general lack of freedom or mobility occasioned by the current situation and its effects thereof. Thus, any desires to return were on hold due to the fears relating to the ongoing conflict and the apparent dangers, as well as the likely re-displacement and risk to life that any attempts at returning might mean. The latter fear was especially strong since some IDPs had actually returned in the post-MOU period of 2002, when there appeared to be relative calm in such places, only to be re-displaced. Confirming this, the EU (2006) indicates that the numbers of people returning to their place of origin were very high during the 2002–2005, although these had decreased with the progressive resumption of hostilities.

Clearly return is not only envisaged by governments and implementing agencies for policy purposes alone but also by IDPs to various intents and purposes. Thus, the motives for which form of return might be desired or promoted are varied depending on the aims. It definitely does not depend on a simple binary of displacement, 'uprootedness' as opposed to returning home. In effect, encouraging people to go 'home' is not as unproblematic as it is often presented, given the contested meanings attached to home, either as a place of nostalgia to which one must necessarily return for life to be normal or the more dynamic view of home as a place where one can succeed (Brun 2003; Tete 2005). IDPs' conceptions about home are complex, multifaceted and informed by various factors that indicate shifting understandings of home under a given situation rather than a linear idea of home as a place of origin to return to, even if the latter is strongly envisaged. In this particular context, factors such as multiple displacement and the sense of impermanence it occasions, people's personal experience of the war, a pervading sense of insecurity, livelihood concerns, a real and/or perceived sense of well-being, resignation and/or a desire to engage, and uncertainty about the future prospects for peace, were all important reasons why IDPs wanted to return or not return. In this respect, the views expressed in the various life history accounts presented below are informative:

> We left Jaffna in 1995 ... I was 12 years old but I still feel Jaffna is my home. I was born there and we had better facilities when we were living there. Although I am living with my parents and siblings, I have some fear and uncertainty. Because of all the movement, I feel this place [RV] is temporary and so I feel Jaffna is my home. I cannot say exactly when I will return but if there is no conflict in the country, I have a hope that things will improve. (LH account of an IDP youth on sentiments about home and return)

> I was born in Kandy but moved to Killonochi ... Now I feel quite settled in this RV and, especially for my children, this is home even

though I cannot say we're very safe here. My small amount of work in the paddy fields is unreliable, and not always safe to go to, and I'm not satisfied but for the sake of my children. I have to work. It is very difficult for me without my husband but the god protects me. (LH of an IDP widow on her livelihood situation and perception of home)

I am from Omanthai . . . I felt that wherever we were given land, I would go, so far as we were given land . . . For me, Omanthai is home, but now I have to stay here (RV). In Omanthai we had our land, fields, and lots of assets, so I think it is home. If the situation improves, I would still like to go there. We have 7 acres [2.8 ha] of land there, but it is now occupied by the LTTE on one side and the army on another side . . . For me, the future will be for us to return but I don't think the situation will improve now. I want to return but I am afraid for my children. (LH of an IDP woman on her reflections about home, return and the future)

As indicated in the quotes above, several reasons are highlighted as behind IDPs' perceptions about home and return. Those who experienced multiple displacements before coming to the RVs seem to have a strong sense of impermanence which, coupled with a sense of insecurity, tend to make them envisage home as the place of origin or the place from which they were initially displaced. These sentiments about return are also almost invariably attached to a real and/or perceived sense of well-being and the better opportunities they had in the place of origin, and a level of livelihood security which they are missing in the current situation. Related to the latter, there seems to be a strong attachment to land and/or various other assets and the prospects they offer. Thus, for many, home is where there is land, and even though this might be the case in the RVs, its limited extent, coupled with the sense of insecurity, seems to make the desire to return home – to the place where there were more assets – even stronger. Whilst these could be dismissed as idealised, nostalgic images or sentiments of home, they are nonetheless very important notions influencing IDPs' thinking and conceptions about what they consider to be viable solutions.

In the current situation, however, they seem to reckon that fulfilling such wishes is not without risks, and can thus only remain as dreams to be held on to. In fact, many expressed the impossibility of such an undertaking of return, especially while the war is full-scale (see for example BBC News 2008a; 2008b; *ColomboPage* 2008; *DailyMirror* 2008; IASC 2008b for post-fieldwork updates on the escalating war, with Vavuniya especially affected). Yet many seem to be keeping the hope of return alive, even if it remains just a desire. As the IDPs expressed and as was the case during the MOU-CFA period of respite, many did indeed return, if only to be re-displaced. This might be a strong indication that return is not merely a nostalgic ideal but one that IDPs might well be prepared to embark on if the situation was to improve, For many, then, such hopes can only remain a dreams to hold on to since they are not quite sure when and if the saga of 'war for peace' will ever end. Whether they will be able to return to what they think they left behind in this era of occupation, mine-planting, 'cleared' and 'uncleared' politics, and so forth, is another issue altogether and one that is beyond the scope of this discussion. However, giving recognition that IDPs might or would want to return and hence continue to hold on to such hopes rather than giving up their dreams is the least one can do to respect their dignity. At the same time, it provides no excuse for insisting on their 'right to return' under present conditions and without a sustained peaceful solution or stop to the ongoing conflict. This is especially important since what IDPs envisage returning to is not a conflict-ridden, despoiled home stripped of all markers of belonging and beyond recognition, but a home that offers a sense of safety and security, a mooring or anchor, and a place of familiarity not impermanence in the face of uncertainty that is common in the context of raging conflict.

On the other hand, and whilst return might remain an ideal to be achieved, some IDPs have demonstrated a readiness to engage in the process of making a life, especially in the RVs they are living in. The challenge then, remains the promotion of conditions conducive to such an undertaking – conditions that are seriously impeded by the prevailing situation of accelerated fighting and the widespread insecurity it poses, not only to relocation but also to resettlement or return of IDPs in the conflict-affected north-east and adjacent areas of Sri Lanka (see Jayasinghe 2005;[11] Kälin 2007).

## Conclusions – 'In the present prevailing situation, [we are] unable to think'

Brun succinctly notes that,

[O]ne problem with much of the debate on when internal displacement ends is that it is largely focused on the policy point . . . the debate only to a very limited extent takes into consideration the *experiences* of the internally displaced themselves. (2005, 14; my italics)

Although discussing return in the context of refugee repatriations, Koser & Black's (1999, 7, 9, 10) observations are noteworthy: 'indeed in some . . . repatriations, a return to place of origin can be seen as desirable, but not possible. The challenge to policy makers is to incorporate in repatriation initiatives refugees' own meanings, perceptions and expectations of "home"'. Evidently, the discourse on the 'right to return' is fed by a medley of issues, not least of which are essentialist but persisting conceptions of home, geo-politically powerful policy narratives with attendant practices and also some displaced persons' own desires to return.

In the face of the worsening conflict and current context of generalised insecurity, it appears that such solutions can only be promoted for reasons other than the interests of the people on whose behalf they are claimed to be beneficial. Obviously, the argument is not about whether return, or any solution, is desired or desirable as it is about its expediency and the overriding need of IDPs for protection. For any claims to be made with regards to providing a solution of some sort then, it is obvious that efforts have to be directed, not at pushing ahead with the implementation of a set of prescribed 'durable solutions', but at helping address the security situation, without which the very basics of voluntariness, safety and dignity would be seriously undermined. The exercise of all other rights is of necessity subject to the basic *right to life* if one is to lay claim to the enjoyment of any rights at all. I argue that while return may be desired or even preferred, it simply may not be possible under the prevailing circumstances, thereby requiring alternative solutions to the displacement situation.

In giving the word to forced migrants as agency-endowed people, we could then think of them as purposive actors embedded in particular social, political and historical situations –but also as ones who resourcefully interpret and reinterpret the structures, rules and situations that might act to constrain and/or enable them (Long & Long 1992; Shanmugaratnam et al. 2003). Obviously, a careful reading of their voices puts into question the relevancy of focusing on the technical implementation of any particular solutions without taking into account the lived realities of IDPs on the ground. Recognising them as capable and knowledgeable social actors endowed with agency requires that their voices be heard on issues of concern to them, the ways they are acting and specifically, how they perceive the right to return and/or the solutions they consider relevant in the current circumstances. That this is necessary was reiterated by the UN Representative on the Human Rights of IDPs during his mission visit to Sri Lanka when he noted that, '*increased information sharing and consultation* by the government *with* both *the IDPs themselves* and the International Community and agencies, *would substantially contribute to reduce the prevailing sense of insecurity* and facilitate reintegration into areas of return' (Kälin 2007; my italics). It is in this spirit that by way of conclusion, I redirect attention to and encourage a careful reading of the excerpts from the life history account of a conflict-displaced Tamil IDP as to the way forward (Box 1).

Hopefully, the message conveyed in the opening statements and expressed by one, of many, whose experiences should count in the implementation of any solution, is clear enough to merit serious attention from all who have kindly taken it upon themselves and/or claim to have the interest of such internally displaced persons at heart. After all, policy discourses and practices are there to serve the needs of those for whom they are designed. This affords a great opportunity to move beyond mere words to actually listening and incorporating such candid views.

## Notes

1 The GPs were formally proposed by Dr Francis Deng, the former representative of the UN Secretary-General on IDPs, in collaboration with a team of international legal experts, prominent among whom was Walter Kälin, the current representative (see Kälin 2000, v, Preface by Francis Deng).

2 See Kälin (2007, 3) for what a framework for durable solutions expects of governments or is intended to do.

3 Accurate figures as to the actual civilian death toll are not readily available. The estimates range from 65,000 to 75,000, but the number could be higher.

4 UNHCR document 'IDPs by Place of Displacement and Place of Origin as at 31st January 2008' (IDP movements after 7 April 2006), UNHCR/GIS Unit, Sri Lanka.

5 In Sri Lanka, areas under government control, especially those reclaimed from the LTTE in the north-east are and commonly referred to by the GoSL as 'cleared' areas whilst those under LTTE control are considered 'uncleared' areas.

6 It does not fall within the scope of the present paper to discuss the implications of methodological and ethical issues, but a critical reading of the paper is recommended as the information was necessarily framed by the prevailing circumstances.

7 Whilst my reception was generally warm, reactions ranged from shock, to incredulity, to curiosity bordering on exoticism, and in some cases, admiration, especially from the young women as to why I would risk conducting research in their environment, even though they felt encouraged by our presence. I cannot rule out suspicion and even maybe dislike, although I was not openly confronted by such attitudes.

8 My main interpreter was the programme coordinator of the NGO who had also followed a course at the Department of Geography at NTNU. He was instrumental in arranging for us to visit a location which many considered too risky to visit in the line of duty, much less to attempt to conduct research there. Other staff interpreted on a few other occasions.

9 Displacement covers a wide range of situations but in this paper it primarily refers to conflict-induced displacement, specifically experienced by IDPs, who are alternatively referred to as 'forced migrants'.

10 Defined by Bhabha (1996: 50) as 'oppressed, minority groups whose presence was crucial to the self-definition of the majority group: subaltern social groups were also in a position to subvert the authority of those who had hegemonic power'.

11 Mr M.S. Jayasinghe, as Secretary of the Ministry of Relief, Rehabilitation and Reconstruction (MRRR), lists among the Ministry's functions: 'to resettle and relocate IDPs in the North-East and adjacent areas'. He also indicates that 'rehabilitation assistance is to be made available *only* to families who have returned to land and property they own, or which has been granted by the authorities ... and where the *returnee* and his family *have demonstrated a clear intention to remain* in the location *in a permanent capacity*' (Jayasinghe 2005, 8, 12, my italics).

*Acknowledgements.* – I wish to thank Piers Blaikie and Ragnhild Lund for their highly constructive comments on this paper. I also thank Catherine Brun, Hans Skotte, Sarah Khasalamwa, John Nigel, Chamilla Attanapola, and Anders Aspaugh for their valuable comments on all or parts of earlier drafts of the paper. Radmil Popovic is thanked for technical assistance and Catriona Turner for editorial help. Many thanks are due to all the research participants, without whom there would be no story to tell, and also to FORUT-Sri Lanka and FORUT-Vavuniya. P.C. Sivapalan deserves special mention for excellent interpretation services. I also extend my gratitude to two anonymous reviewers for their insightful comments.

## References

Agterhuis, S. 2005. *The Right to Return and Its Practical Application.* Aristotle University of Thessaloniki and Palestinian Refugee ResearchNet (PRRN), Montreal. http://www.arts.mcgill.ca/mepp/new_prrn/research/papers/agterhuis.pdf (accessed January 2008).

Alarcon, N. 1994. Traddutora, traditora: A paradigmatic figure of Chicana feminism. Grewal, I. & Kaplan, C. (eds.) *Scattered Hegemonies. Postmodernity and Transnational Feminist Practices*, 110–133. University of Minnesota Press, Minneapolis.

Amnesty International. 2001. *Israel and the Occupied Territories/Palestinian Authority: The Right to Return: The Case of the Palestinians.* Policy Statement of Amnesty International. http://www.amnesty.org/en/library/asset/MDE15/013/2001/748cdbc1-a4e0-11dc-a92d-271514ed133d/mde150132001en.html (accessed January 2008).

Amnesty International. 2006. *Sri Lanka: Waiting to Go Home – The Plight of the Internally Displaced.* Amnesty International Document, June 2006. http://www.amnesty.org/en/library/info/ASA37/004/2006 (accessed March 2008).

Banerjee, P. 2006. IDP protection at the national level in South Asia. *Forced Migration Review (FMR)*, Special Issue, December 2006, pp. 18–20. http://www.fmreview.org/FMRpdfs/BrookingsSpecial/full.pdf (accessed October 2008).

BBC News. 2008a. *A Grim Future Foretold for Sri Lanka.* 8 January 2008. http://news.bbc.co.uk/2/hi/south_asia/7176236.stm (accessed September 2008).

BBC News. 2008b. Indians Hurt in Sri Lanka Clash. 11 September 2008. http://news.bbc.co.uk/2/hi/south_asia/7610704.stm (accessed September 2008).

Bhabha, H.K. 1996. Unsatisfied: Notes on vernacular cosmopolitanism. Garcia-Moreno, L. & Pfeiffer, P.C. (eds.) *Text and Nation: Cross-disciplinary Essays on Cultural and National Identities*, 191–207. Camden House, Columbia, SC.

Blaikie, P.M., Springate-Baginski, O., Banerjee, A., Bhatta, B., Saigal, S. & Sarin, M. 2007. Actors and their narratives in participatory forest management. Springate-Baginski, O. & Blaikie, P.M. (eds.) *Forests, People and Power: The Political Ecology of Reform in South Asia*, 93–115. Earthscan, London and Sterling, VA.

Bradley, M. 2007. *Return in Dignity: A Neglected Protection Challenge*. RSC Working Paper No. 40. Refugee Studies Centre, Oxford. http://www.rsc.ox.ac.uk/PDFs/RSCworkingpaper40.pdf (accessed January 2008).

Brookings-Bern. 2007. *When Displacement Ends: A Framework for Durable Solutions*. The Brookings Institution – University of Bern Project on Internal Displacement, June 2007. http://www.brookings.edu/ ~/media/Files/rc/reports/2007/09displacementends/2007_durablesolutions.pdf (accessed October 2008).

Brun, C. 2003. *Finding a Place. Local Integration and Protracted Displacement in Sri Lanka*. Dr.Polit Thesis, Department of Geography, Norwegian University of Science and Technology, Trondheim.

Brun, C. 2005. *FMO Research Guide: Internal Displacement*. Vol. 41. http://www.forcedmigration.org/guides/fmo041/ (accessed January 2008).

Clarance, W. 2007. *Ethnic Warfare in Sri Lanka and the UN Crisis*. Pluto Press, London and Vijitha Yapa Publications, Colombo.

*ColomboPage*. 2008. *LTTE launches another artillery attack on the Vavuniya Air Force base*. 9 September 2008. http://www.colombopage.com/archive_08/September9163920RA.html (accessed September 2008).

CPA (Centre for Policy Alternatives). 2001. *Defending the Rights of Internally Displaced Persons*. http://www.cpalanka.org/IDP_project.html (accessed January 2008).

Creswell, J. 2003. Research Design: Qualitative, Quantitative and Mixed Method Approaches. Sage, Thousand Oaks, CA.

*DailyMirror*. 2008. Opinion: Did a foreigner man the LTTE aircraft? Army repulses major attack on military complex. 12 September 2008. http://www.dailymirror.lk/DM_BLOG/Sections/frmNewsDetailView.aspx?ARTID =26093 (accessed September 2008).

EU (European Commission). 2006. *Humanitarian Aid for People Displaced by the Conflict in Sri Lanka*. Humanitarian Aid Decision 23 02 01. European Commission Directorate General for Humanitarian Aid – ECHO, December 2006, Brussels. http://www.ec.europa.eu/echo/pdf_files/decisions/2006/dec_srilanka_01000.pdf (accessed January 2008).

Ferris, E. 2007. Durable solutions: When does displacement end? *IDP Newsletter*. Consortium of Humanitarian Agencies, Brookings Institute. http://www.brookings.edu/articles/2007/12_displacement_ferris.aspx (accessed March 2008).

Haraway, D. 1991. *Simians, Cyborgs, and Women: The Reinvention of Nature*. Routledge, New York.

Hyndman, J. 2000. *Managing Displacement. Refugees and the Politics of Humanitarianism*. University of Minnesota Press, Minneapolis.

Hyndman, J. & de Alwis, M. 2004. Bodies, shrines, and roads: Violence, (im) mobility and displacement in Sri Lanka. *Gender, Place and Culture* 11:4, 535–557.

IASC (Inter-Agency Standing Committee). 2008a. Country Team Situation Report #112 on Jaffna, Kilinochchi, Mullaitivu, Mannar, Vavuniya, Trincomalee, Batticaloa and Ampara Districts. 31 January – 7 February 2008. Colombo, Sri Lanka. http://www.reliefweb.int/rw/RWFiles2008.nsf/FilesByRWDocUnidFilename/MUMA-7BR9T4-full_report.pdf/$File/full_report.pdf (accessed April 2008).

IASC (Inter-Agency Standing Committee). 2008b. *Situation Report #140. Jaffna, Kilinochchi, Mullaitivu, Mannar, Vavuniya, Trincomalee, Batticaloa and Ampara Districts*. 14–21 August 2008. Colombo, Sri Lanka. http://www.reliefweb.int/rw/RWFiles2008.nsf/FilesByRWDocUnidFilename/LSGZ-7HUHDU-full_report.pdf/$File/full_report.pdf (accessed September 2008).

ICRC (International Committee of the Red Cross). 2008. *ICRC in Sri Lanka*. Press Briefing. http://www.cicr.org/Web/Eng/siteeng0.nsf/html/sri_lanka!Open#Press%20briefing (accessed March 2008).

IDMC (Internal Displacement Monitoring Centre). 2006. *Global Statistics: IDP Country Figures*. Internal Displacement Monitoring Centre and Norwegian Refugee Council, Geneva, Switzerland. http://www.internal-displacement.org/8025708F004CE90B/(httpPages)/22FB1D4E2B196DAA802570BB005E787C?OpenDocument&count=1000 (accessed March 2007).

IDMC (Internal Displacement Monitoring Centre). 2007. *Civilians in the Way of Conflict: Displaced People in Sri Lanka*. Internal Displacement Monitoring Centre and Norwegian Refugee Council, Geneva, Switzerland. http://www.internal-displacement.org/8025708F004BE3B1/(httpInfoFiles)/882EB0DAEEA545A4C12573620026E7EB/$file/Sri_lanka_special_report_sep07.pdf (accessed March 2007).

IDMC (Internal Displacement Monitoring Centre). 2008. *Internal Displacement: Global Overview of Trends and Developments in 2007*. April 2008. NRC (Norwegian Refugee Council)/IDMC (Internal Displacement Monitoring Centre), Geneva. http://www.internal-displacement.org/8025708F004BE3B1/(httpInfoFiles)/BD8316FAB5984142C125742E0033180B/$file/IDMC_Internal_Displacement_Global_Overview_2007.pdf (accessed September 2008).

IRIN (Integrated Regional Information Network). 2002. *In-depth: Internal Displacement. Africa: Introduction*. http://www.irinnews.org/IndepthMain.aspx?IndepthId=41&ReportId=70979 (accessed May 2008).

IRIN (Integrated Regional Information Network). 2008. *Sri Lanka: Half a Million People Could Be Affected by Conflict in 2008*. IRIN Asia, Sri Lanka. February 2008. http://www.irinnews.org/Report.aspx?ReportId=76930 (accessed March 2008).

Jacobsen, K. & Landau, L.B. 2003. The dual imperative in refugee research. Some methodological and ethical considerations in social science research on forced migration. *Disasters* 27:3, 185–206. DOI:10.1111/1467-7717.00228.

Jayasinghe, M.S. 2005. *Rehabilitation and Reconstruction Activities in North and East Provinces*. Development Forum 2005 – Background Papers. http://www.erd.gov.lk/devforum/Treasury/paper-2.pdf (accessed January 2008).

Jayasundara, R. 2007. Settling where displaced: Dependency on government cripples Poonthottam camp. *IDP Newsletter* 2:1, 9. January 2007. http://www.humanitarian-srilanka.org/CKC/DailyInfo/IDP%20News%20Letter%20January%20February.pdf (accessed January 2008).

Johnston, I. 2008 *Medea*. Translated by I. Johnston from the version written by Euripides in 431 BC. http://malaspina.edu/~johnstoi/euripides/medea.htm (accessed January 2008).

Kälin, W. 2000. *Guiding Principles on Internal Displacement: Annotations*. Studies in Transnational Legal Policy No. 32. The American Society of International Law and The Brookings Institution Project on Internal Displacement, Washington DC.

Kälin, W. 2001. *How Hard is Soft Law? The Guiding Principles on Internal Displacement and the Need for a Normative Framework*. Presentation at Roundtable Meeting, Ralph Bunche Institute for International Studies, CUNY Graduate Center, 19 December 2001. http://www.brookings.edu/fp/projects/idp/articles/Kaelin12-19-01.pdf (accessed May 2008).

Kälin, W. 2007. *UN Expert Emphasizes Sustainable and Durable Solutions for Sri Lanka's Internally Displaced Persons*. UN Press Release. Brookings-Bern Project on Internal Displacement. http://www.brookings.edu/projects/idp/RSG-Press-Releases/20071227_sri_lanka.aspx (accessed January 2008).

Kibreab, G. 1999. Revisiting the debate on people, place, identity and displacement. *Journal of Refugee Studies* 12:4, 384–410.

Koser, K. & Black, R. 1999. The end of the refugee cycle? Black, R. & Koser, K. (eds.) *The End of the Refugee Cycle? Refugee Repatriation and Reconstruction*, 2–17. Berghahn Books, New York and Oxford.

Limb, M. & Dwyer, C. (eds.) 2001. *Qualitative Methodology for Geographers. Issues and Debates*. Arnold, London and Oxford University Press, New York.

Liu, G. 2008. The right to return in China. *International Migration* 46:1, 191–231, Blackwell Publishing. http://www.ingentaconnect.com/content/bpl/imig (accessed January 2008).

Long, N. & Long, A. (eds.) 1992. *Battlefields of Knowledge: The Interlocking of Theory and Practice in Social Research and Development*. Routledge, London and New York.

Lund, R. 2003. Representations of forced migration in conflicting spaces: Displacement of the Veddas in Sri Lanka. Shanmugaratnam, N., Lund, R. & Stølen, K.A. (eds.) *In the Maze of Displacement. Conflict, Migration and Change*, 76–104. Høyskoleforlaget – Norwegian Academic Press, Kristiansand.

Malkki, L.H. 1992. National geographic: The rooting of peoples and the territorialisation of national identity among scholars and refugees. *Cultural Anthropology* 7:1, 24–44.

Malkki, L.H. 1997. News and culture: Transitory phenomena and the fieldwork tradition. Gupta, A. & Ferguson, J. (eds.) *Anthropological Locations. Boundaries and Grounds of a Field Science*, 86–101. University of California Press, California.

Martin, S. & Charny, J. 2005. *Sri Lanka: Imperative to Respond to Needs of Conflict-Displaced Refugees.* International Bulletin, 28 October 2005. http://www.refugeesinternational.org/content/article/detail/7140/ (accessed March 2008).

Mooney, E. 2003. Bringing the end into sight for internally displaced persons. *Forced Migration Review (FMR)* 17, 4–7. http://www.fmreview.org/FMRpdfs/FMR17/fmr17full.pdf (accessed January 2008).

Newman, J. 2003. *Narrating Displacement: Oral Histories of Sri Lankan Women.* Refugee Studies Centre (RSC) Working Paper No. 15. International Development Centre, University of Oxford, Oxford.

Nseroko, D.D.N. 1981. The right to return home. *Indian Journal of International Law* 21, 336. http://www.arts.mcgill.ca/mepp/new_prrn/research/papers/agterhuis.pdf (accessed January 2008).

OHCHR (Office of the High Commissioner for Human Rights). 1976. *International Covenant on Civil and Political Rights.* http://www2.ohchr.org/english/law/ccpr.htm (accessed January 2008).

Repstad, P. 1993. *Mellom noerhet og distanse.* Universitetforlaget, Oslo.

Richmond, A.H. 1988. Sociological theories of international migration: The case of refugees. Harrell-Bond, B.E. & Monahan, L. (eds.) *The Sociology of Involuntary Migration, Current Sociology* 36:2 (Special Issue), 7–25.

Richmond, A.H. 1994. *Global Apartheid: Refugees, Racism, and the New World Order.* Oxford University Press, Toronto.

Robinson, J. (ed.) 2002. *Development and Displacement*. The Open University in association with Oxford University Press, Oxford.

Sandlund, A. 2005. EQUITY in the post-tsunami context in Sri Lanka. *IDP Newsletter* 2, 1–3.

Sepulveda, D.C. 1995. Challenging the assumptions of repatriation. *The Courier* 150, 83–85.

Shanmugaratnam, N., Lund, R. & Stølen, K.A. 2003. *In the Maze of Displacement: Conflict, Migration and Change.* Høyskoleforlaget –Norwegian Academic Press, Kristiansand.

Skonhoft, C.G. 1998. *Coping in Adversity. A Study of Burundian and Rwandan Hutu Refugees during Flight and Exile in Tanzania.* Master's thesis, Department of Geography, NTNU, Trondheim.

Spivak, G.C. 1988. Can the subaltern speak? Nelson, C. & Lawrence, G. (eds.) *Marxism and the Interpretation of Culture*, 271–313. University of Illinois Press, Urbana, IL.

Tete, S.Y.A. 2005. *Narratives of Hope? Displacement Narratives of Liberian Refugee Women and Children in the Gomoa-Budumburam Refugee Camp in Ghana*. Master's thesis in Development Studies, Department of Geography, NTNU, Trondheim.

Thompson, P. 1981. Life histories and the analysis of social change. Bertaux, D. (ed.) *Biography and Society: The Life History Approach in the Social Sciences*, 289–306. Sage, Beverly Hills, CA.

Turton, D. 2002. Forced displacement and the nation-state. Robinson, J. (ed.) *Development and Displacement*, 19–75. Oxford University Press, Oxford, in association with The Open University, Milton Keynes.

Turton, D. 2003. *Conceptualising Forced Migration*. Refugee Studies Centre (RSC) Working Paper No. 12. International Development Centre, University of Oxford, Oxford.

UN (United Nations). 1948. *The Universal Declaration of Human Rights.* http://www.un.org/Overview/rights.html (accessed January 2008).

UNHCR. 1998. Report of the Representative of the Secretary-General, Mr. Francis M. Deng, Submitted Pursuant to Commission Resolution 1997/39. Addendum: Guiding Principles on Internal Displacement. 11 February 1998. E/CN.4/1998/53/Add.2. UN Commission on Human Rights, Geneva.

UNHCR. 2003. The road to recovery. Sri Lanka: emerging from the ruins. *Refugees* 1:130, 5–22.

USDS (US Department of State). 2007. *Background Note: Sri Lanka, Bureau of South and Central Asian Affairs.* http://www.state.gov/r/pa/ei/bgn/5249.htm (accessed March 2008).

Uyangoda, J. 2007. Ethnic conflict in Sri Lanka: Changing dynamics. *Policy Studies* 32. East-West Centre, Washington.

Valentine, G. 2001. At the drawing board: Developing a research design. Limb, M. & Dwyer, C. (eds.) *Qualitative Methodology for Geographers. Issues and Debates*, 41–54. Arnold, London and Oxford University Press, New York.

Van Hear, N. & Rajasingham-Senanayake, D. 2006. From complex displacement to fragile peace in Sri Lanka. Van Hear, N. & McDowell, C. (eds.) *Catching Fire: Containing Forced Migration in a Volatile World*, 44–71. Lexington, Oxford and Rowman & Littlefield, Lanham, MA.

Walters, W. 2004. Secure borders, safe haven, domopolitics. *Citizenship Studies* 8:3, 237–260.

WFP (World Food Programme). 2000. *Review of the Protracted Relief and Recovery Operation, Sri Lanka 6152*. WFP, Colombo.

# 'Tsunami third wave' and the politics of disaster management in Sri Lanka

KALINGA TUDOR SILVA

The paper compares the social and political consequences of two disasters that struck Sri Lanka 70 years apart, the 1934–1935 malaria epidemic that killed over 100,000 people and the tsunami of 26 December 2004. The main argument is that while the state and civil society responses to the malaria epidemic of 1930s in many ways led to the establishment and development of the Sri Lankan welfare state and the related outcomes, including rapid advances in quality of life particularly in the rural sector, the more elaborate and well-funded tsunami response driven by the international humanitarian industry failed to facilitate speedy recovery, failed to galvanize the peace process and even added to the vulnerability of some of the affected people due to the rapid exodus of many newly arrived INGOs before completing their humanitarian mission. While the relative failure of the tsunami response may be partly attributed to the persistent conflict between the Government of Sri Lanka and the Liberation Tigers of Tamil Eelaam (LTTE), it also exposes some of the key weaknesses of humanitarian aid in the modern world. A more historically grounded and conflict-sensitive approach is necessary along with greater coordination among agencies involved.

*Kalinga Tudor Silva, Department of Sociology, University of Peradeniya*

## Introduction

Some commentators have characterized the tsunami of 26 December 2004 as the worst natural disaster ever recorded in the known history of modern Sri Lanka (de Silva 2005).[1] Many observers do not realize that the malaria epidemic of 1934–1935 was far more devastating in terms of mortality caused, the official death toll attributed to the epidemic and the accompanying famine being 80,000 as compared to the official death toll of 40,000 attributed to the tsunami. While the tsunami devastated the coastal belts of eastern and southern Sri Lanka within the space of less than one hour, the 1934–1935 epidemic unfolded over nearly 15 months across central Sri Lanka with debilitating long-term consequences. It is important to point out that the 1934–1935 epidemic played an important role in the development of civil society, the emergence of the Sri Lankan welfare state and the formation of key players of this welfare state, including nationalist and leftist leaders. Sri Lanka has experienced a spate of man-made and natural disasters in more recent years, including civil war, floods, landslides, and cyclones. A more historically grounded understanding of these disasters and the relevant responses from state, civil society and international agencies is needed in order to, firstly, assess their overall impact on society, culture and patterns of development or underdevelopment, and secondly, to evolve suitable humanitarian and developmental responses that build on local strengths, capacities and adaptations. This paper seeks to arrive at a historically informed understanding of the impact of and response to natural disasters, paying special attention to the outcomes of the 1934–1935 epidemic and the tsunami of 2004.

Unlike the malaria epidemic of 1934–1935, which received considerable publicity only in Sri Lanka and, to a limited extent in Britain, the tsunami of 26 December 2004 was a uniquely globalized event, closely and instantly watched through the global electronic media worldwide, with a resulting humanitarian response of an unprecedented magnitude. The 1934–1935 disaster was due to an unfortunate conflation of natural disaster (drought) and an epidemic disease (malaria). The progressive neglect of the Sri Lankan peasantry by the colonial state itself was an important factor contributing to the epidemic and the famine associated with it. Many members of the Sri Lankan elite became fully aware of poverty and deprivation in the countryside at the time through the widely publicized effects of this disaster and this, in turn, shaped the development of a range of civil society organizations responding to the relevant needs. More importantly, the Sri Lankan political leadership in the last years of colonial rule responded to the situation by formation of a welfare state, with universal free health care, free education, and the establishment of peasant colonization schemes as key pillars of this welfare state. Many of Sri Lanka's subsequent gains in quality of life can be attributed to the policy regimes that evolved in response to the transfer of power from the British Raj to Sri Lankan leaders and the emergence of electoral politics under the Donomore Constitution of 1931. In contrast, the tsunami of 2004 occurred at a time of erosion of the welfare state and serious contestation of the power of the Sinhala-dominated welfare state, particularly in some of the tsunami-affected areas with a preponderance of minority ethnic groups. As a complex humanitarian emergency of the modern era the tsunami led to a massive influx of donor support in the aftermath of the event, but much of it and the optimism that came with it had receded by 2007, causing widespread disenchantment and disappointment among the tsunami victims, as captured by the term 'third wave' (the first two waves being tsunami waves and the third and most devastating of all being the wave of tsunami aid that ended abruptly). In the tsunami response in Sri Lanka, this situation evolved due to a complex web of factors, including unresolved conflicts between the state and the Liberation Tigers of Tamil Eelaam (LTTE), disagreements between the state and NGO (national government organization) community over a range of issues, the politics of international aid itself and an acute dependency syndrome generated among the beneficiaries of international aid.

The primary analytical question posed in this paper is as follows. If the serious malaria epidemic of the 1930s that occurred in a colonial but decolonizing setting led to important endogenous developments and responses such as the development of a civil society and formation of a welfare state, why did an even more spectacular disaster such as the tsunami, which occurred in a far more globalized and a post-colonial setting, fail to galvanize similar lasting societal responses from within Sri Lanka?

This, in turn, points to the need for a re-examination of the current paradigms of relief, recovery and rehabilitation guiding the global humanitarian responses. By comparing responses to the malaria epidemic of 1934–1935 and the tsunami of 26 December 2004, this paper seeks to identify historical continuities and ruptures in disaster response in Sri Lanka. A full answer to this analytical question calls for more detailed contextual and historical research than is feasible in the present paper, but it prepares the ground for such detailed research.

## The malaria epidemic of 1934–1935

The great malaria epidemic of 1934–1935 was an important landmark in the history of malaria in Sri Lanka. In the words of Gill (1936, 1), an official chronicler of the epidemic, it was 'the greatest pestilence in the recorded history of the island'. H.E. Newman, who was in charge of the relief effort, recorded it as 'the greatest material disaster to befall the island since the coffee smash of the late nineteenth century' (Newman 1935, cited in Jones 2000). According to official estimates it wiped out 80,000 people or *c.*1% of the total population of the country at the time. According to some estimates almost half of the population in the country became ill with malaria during this epidemic, which lasted for a period of eight months (Briercliffe 1935; Gill 1935). There were two waves of the epidemic, one from October 1934 to March 1935 and the other from April to August 1935. The worst affected areas were the Districts of Kegalle and Kandy in the Wet Zone and the adjoining District of Kurunagala in the Intermediate Zone, predominantly inhabited by the Sinhalese peasants. However, since this epidemic had countrywide impact malaria entered the consciousness and social and political agendas of the elite in the country for the first time (Meyer 1982). Not surprisingly, the epidemic attracted worldwide attention in tropical medicine and malariology (e.g. Harrison 1978). This major catastrophe was a key factor that shaped Sri Lanka's early adoption of a malaria eradication strategy. Another aspect of the epidemic was that besides evoking an official response from the colonial state, it also evoked a social and political response from within the country (Silva 1994, 1997).

The official investigation instituted by the state utilized a range of vital statistics, entomological and parasitological data, weather information, hospital records, and field visits to determine the causes and impact of the epidemic (Briercliffe 1935; Gill 1936). The official explanation of the epidemic centred on two fronts, one climatological and the other immunological. According to the climatological argument, Sri Lanka in general, and the affected areas in particular, had experienced prolonged drought in the preceding two years and, as a result, rivers in the affected areas had become reduced to stretches of shallow pools exposed to sunlight, giving rise to an environment highly favourable for the rapid breeding of *Anopheles culicifacies* (mosquito) along the riverbeds.

The immunological explanation stressed the lack of immunity against malaria in the wet zone population following a period of good health in the preceding years. Given a cyclical pattern of malaria epidemics occurring at roughly five-year intervals, with previous epidemic outbreaks reported in 1906, 1911, 1914, 1919, 1923, and 1928–1929, the possibility of malaria relapses acting upon a non-immune population was pointed out. In accounting for this massive epidemic these climatological and entomological explanations and immunological arguments were buttressed with brief references to the economic stresses at the time (Silva 1994; 1997). Mention was made of the worldwide depression of the 1930s preceding the epidemic and the resulting food scarcities in the country aggravated by crop failure resulting from the prolonged drought. The Sinhala newspapers of the era commonly identified famine (*durbiksha*) as a characteristic feature of the epidemic.

### *The impact of the epidemic*

The epidemic had the greatest impact in Sinhala rural communities in interior parts of Sri Lanka. Many of these communities earned their livelihoods from smallholder agriculture and wage labour but they were poorly integrated with the plantation economy established in neighbouring areas by European and local planters. As Meyer (1982) discovered, some of the worst affected were so-called depressed caste communities in Rambukkana, Kegalle and Kurunagala areas. The official investigations revealed that in view of their malnutrition and lower immunity levels children suffered most.

The infant mortality rates (IMR) in the relevant districts increased substantially during the epidemic period, in some cases by three or four times compared to the pre-epidemic period (Table 1).

There were, however, far more moving first-hand accounts of the suffering of the affected people. F.C. Gimson, the Assistant Government Agent of Kegalle, after house to house visits in his area, reported:

> [I]n the first house I visited, I found a woman had just given birth to a child and no one was in a fit state to attend to its wants as the rest of the family was stricken with fever. The village was two miles from the road, across a swift flowing river, and it was impossible for the mother of the child to be taken to the hospital. . . . both were dying . . . In the next house everyone was down with fever with the exception of a small child [who was] eighteen months old. She was sprawling about, howling for food which her mother in a state of delirium was unable to give. (Jones 2000, 95)

In State Council debates the representatives from malaria-affected constituencies reported various horror stories after

*Table 1.* Infant mortality rates (IMR) in the epidemic districts during and prior to the epidemic

| Epidemic district | IMR 1923 – 1932 average | IMR for last quarter of 1934 and first quarter of 1935 |
|---|---|---|
| Colombo | 148 | 196 |
| Negombo | 135 | 270 |
| Kalutara | 117 | 128 |
| Kandy | 184 | 310 |
| Matale | 216 | 497 |
| Kurunagala | 233 | 800 |
| Chilaw | 159 | 311 |
| Ratnapura | 161 | 233 |
| Kegalle | 138 | 452 |

*Source:* Briercliff (1935, 44).

visiting their constituencies. Sir D.B. Jayatilaka, Leader of the House, reported:

> I find during the course of my visits to a number of villages and dispensaries that it is the children who are the principle sufferers from malaria. Even when the parents themselves are unable to work on account of fever they have to stay at home to look after the children. This absence from work seriously affects those who are dependent for their livelihood on daily paid labour ... These villagers are at present well-nigh destitute and though it is possible that they might obtain rice at the local boutique on credit, or borrow from their neighbours, their children are not able to digest this form of diet and a relapse results. (Government of Ceylon 1935, 3158)

A. Ratnayake, who represented the Dumbara constituency that was badly affected by the epidemic, reported during his submissions in the State Council that he had visited a village where,

> In every home practically every inmate is ill. ... I saw with my own eyes the state of affairs in a certain homestead ... where there were six individuals living in two small rooms; one patient was dying, and the others were too ill to be able to do anything ... the family was face to face with starvation. (Government of Ceylon 1935, 3051)

The accounts in the popular Sinhala newspapers were even more moving. On 14 June 1935, the popular Sinhala daily newspaper, *Dinamina* reported on a man ill with fever having to carry the dead body of his child in one hand and a *mamothy* (an implement used for digging the ground) in the other hand in order to bury the child with his own labour as nobody else in the village could assist him in this pathetic moment as they were all suffering from malaria. On 21 December 1934, the Dinamina carried a story about a man who had committed suicide because he had been unable to bear the sudden death of his child following a malaria attack.

## *The role of the state*

It is significant that the epidemic broke out subsequent to the introduction of universal franchise and establishment of gradual self-rule in Sri Lanka since 1931. As a result the state had to take serious note of the epidemic, and consequently, it played a pivotal role in providing emergency medical services and relief to the affected people.

## *Emergency medical services*

As regards the emergency medical services, initially the health authorities sent out itinerant medical officers to affected villages to distribute free quinine. Soon this was found to be impractical given the large number of sick persons in each area. Then a temporary treatment centre was established for a group of affected villages, ensuring that such a centre existed within three miles of affected villages. By January 1935 there were 68 government hospitals, 95 government central dispensaries, and 710 branch dispensaries and temporary treatment centres in the affected areas (Jones 2000, 96). In addition, many medical practitioners and medical students from other areas were temporarily mobilized in the emergency medical service. The sanitary inspectors were mobilized to collect first-hand information as to the number of people affected and whether they were receiving the services. The state expenditure on medical services increased from 9.7% of the total revenue of the state in 1931 to 11.1% in 1935 (Jones 2000, 95). The estate health services provided treatment to the malaria patients from nearby villages in addition to serving the estate residents who were the legitimate beneficiaries of the estate medical services.

There are many reports of the dedication with which medical staff provided their services. In State Council debates mention was made of one doctor in Kurunagala who had seen 5000 patients in five days. The medical staff worked for long hours, sometimes more than 12 hours per day. Many health workers themselves became infected with malaria. At least one doctor died of malaria while working in an epidemic area.

There were, however, many shortcomings in the system due to overcrowding, shortages of drugs, and transport difficulties faced by patients in remote areas. People were travelling on foot and bullock carts to access medical treatment at the nearest health outposts and had to wait long hours before they could see a doctor. All observers noted the long queues in the medical centres. *The Times of Ceylon* of 17 December 1934 reported that at the Kurunagala hospital, there were patients 'on the beds, under the beds, on the verandahs – anywhere and every where. Some with neither beds nor mats, lay huddled up on the bare cement ... neither knowing nor caring what was to happen to them' (Jones 2000, 94).

Vector control efforts were also carried out by the state, with priority given to anti-larval measures targeting riverbeds where water pools had been formed due to the drought. The state also planned to introduce legal measures in the form of anti-mosquito legislation recommended by some malaria experts, but the State Council did not support these legal measures arguing that it would add to the burdens of the victims of the disaster.

## *Relief to victims of the epidemic*

Largely due to the pressure from representatives of the affected areas in the State Council, a comprehensive programme of relief was initiated once the impact of the epidemic started to become clear. While the colonial government was initially reluctant to embark on economic relief arguing that it would lead to 'pauperization' and continuous demand for such assistance, it had to give in to pressure from representatives in the State Council and the media to initiate public action catering to the poor and destitute in the affected areas. Newnham was appointed the Special Commissioner for Relief on 15 January 1935 and he began his activities by touring the affected areas (Newnham 1935). The relief efforts consisted of the following measures:

(1) Introduction of a free rice ration to selected beneficiaries identified by headmen in the respective areas
(2) Introduction of a free school meal in affected regions
(3) Provision of free malted milk to undernourished infants recovering from malaria, via hospitals and clinics in the severely affected areas
(4) Food-for-work programmes implemented in selected areas in order to provide relief to those whose livelihoods were affected by malaria.

The effectiveness of these programmes varied and there were many instances of malpractice in the implementation of the relief programme (Meyer 1982). However, the important point is that the state became directly involved in provision of emergency medical treatment and relief in affected populations. In many ways, the beginning of the Sri Lankan welfare state can be traced back to the activities of the state during the 1934–1935 epidemic. The heavy involvement of the state in the disaster response during 1934–1935 prepared the ground for the development of a welfare state in the subsequent era with free health care and free education being the cornerstone of the Sri Lankan welfare state from the 1940s (Jayasuriya 2000).[2] The beneficiaries of the relief work carried out during and after the epidemic were primarily Sinhala peasants in the affected areas, who continued to receive favourable treatment from the welfare state even after the epidemic, creating a kind of inherent limitation of the Sri Lanka welfare state that eventually led to a serious polarization of society along ethnic lines.

### *Mixing politics with quinine*

Civil society was heavily involved in the relief operations during and after the malaria epidemic. Civil society organization that directly participated in the relief efforts included religious organizations, private benefactors and politically oriented groups. H.L. Seneviratne has recorded in his book *The Work of Kings* how Sinhala Buddhist leadership became involved in the relief effort during the epidemic:

> While there was some interest on the part of the government to help the people suffering under the malaria epidemic of 1934, Kalukodayave realized that the government was not capable of providing necessary relief, nor was it just to expect the government to do so, which again reflects an unusually sense of civic responsibility and initiative. He therefore resolved to solicit support from individuals ... As before, he first met the member of the State Council for the area ... and walked through the villages in the Hevagam Korale on a fact finding mission. The facts he found were deeply disturbing ... He had successfully collected enough material and medication to deal with the problem in his own area, Hevagam Korale. But the news from Kegalle was bad. He then joined the All Ceylon Buddhist Congress in an effort to collect and transport the needed supplies. He preached in the night to solicit help for the sick and used his radio sermons for the same purpose. ... then help poured in. Many carried the needed supplies to the villages themselves. It is as if Boddhisatva had appeared all over the land. (Seneviratne 1999, 77)

There were many forms of voluntary helps for malaria victims. Some organizations set up temporary health services in the epidemic zone, while others provided generalized assistance, including dry rations. Among the key actors in the relief work were some leftist leaders who clearly identified the epidemic as an opportunity to reach out to the poor and marginalized.

The 1934–1935 malaria epidemic played a significant role in the evolution of the leftist movement in Sri Lanka. The Suriya Mal Movement of the 1930s, the precursor of the leftist movement in Sri Lanka, gained ground in selected rural areas through its participation in the relief work during the epidemic (Lerski 1969; Meyer 1982). Many of the founders of the leftist movement in Sri Lanka, who received their Marxist indoctrination in universities in England, developed a rural base through their collective response to the epidemic. The Suriya Mal Movement started as a fund-raising campaign for the welfare of the poor and it gradually articulated anti-colonial sentiments. During the epidemic the Suriya Mal Movement became actively involved in relief work under the leadership of Dr N.M. Perera (economist), Dr Colvin R. de Silva (lawyer and historian), Phillip Gunawardena (agriculturalist), and Dr S.A. Wickramasinghe (physician).

In contrast to the scientific explanation of the epidemic giving primacy to ecological factors involved in the epidemic, the leftist leaders held the government entirely responsible for the epidemic, which in their estimate had wiped out *c.*180,000 people. In one of his fiery speeches in the State Council, Phillip Gunawardena accused the state of 'cutting down essential services, especially in matters pertaining to health that made the malaria epidemic so acute ... The hospitals were understaffed, the dispensaries were undermanned and some hospitals had actually been closed' (Lerski 1969, 19).

In contrast to the malariological accounts of the epidemic stressing prolonged drought as its primary cause, the leftist discourse on the event focused on its socio-economic causes. As one historian of the leftist movement noted 'the leftist angry young men were convinced that it was primarily malnutrition, the result of evil colonial exploitation, which caused so many people in the villages to suffer and die' (Lerski 1969, 20). Accordingly, their own efforts concentrated on distribution of food and quinine, especially in the Kegalle District. Initially, suspecting their motives, the state refused to cooperate with them, but as the epidemic progressed there developed a degree of mutual understanding between the two parties to the extent that some state assistance was channelled through the Suriya Mal workers. However, it is interesting that the official reports on the epidemic did not make any mention of their contribution. A malariologist noted sarcastically,

> The dispensaries were surrounded by dense crowds of patients, many of them shaking and vomiting with malaria. At many of the dispensaries were 'side-shows': kitchens at which local philanthropists (and sometimes, it may be added, local politicians) provided free tea, coriander water, or rice *congee*. (Wigglesworth 1936, 614)

For the leftist leaders, the malaria epidemic of 1934–1935 provided an eye-opener to the evils of colonial rule. In 1963 one leftist writer remarked:

> The two things given to each patient were a bottle of the standard quinine mixture and Marmite rolled into the form of *vedarala's* [native doctor's] pills... such was the degree of starvation among the peasantry. The Suriya Mal workers were amazed to see how this little [amount of] Marmite revived them and put some life back into them. (Sybil 1963, 6)

Work in this connection was also an eye-opener to many such people who were just getting to know the peasant masses. The poverty was incredible, the overcrowding even more so, with 15, 20, or more people crammed into tiny huts, dying like flies: 'This was what colonial exploitation meant: worse than the worst that prevailed in England when Marx and Engels analyzed the conditions of the working classes. This was what had to be fought' (Sybil 1963, 4–10).

The leftist intellectuals saw epidemic malaria as yet another manifestation of the basic issues of poverty and inequality in the prevailing social system. The leftist leaders were an important source of criticism against state policies and programmes in relation to the epidemic and they were, in turn, criticized by Sir D.B. Jayatilaka, the Leader of the House in the State Council, for 'mixing politics with quinine' (Government of Ceylon 1935, 4490).

Thus, the malaria epidemic of 1934–1935 contributed in its own way to the evolution of the Sri Lankan welfare state and the development of a civil society supportive of welfarism, which became an important pillar of politics and policy making in Sri Lanka in the time to come. Confirming Amartya Sen's (1982) view that famines do not break out in functioning democracies, events similar to the 1934–1935 epidemic did not recur in Sri Lanka in the period that followed. Due to effective malaria control strategies pursued, as well as to the effectiveness of welfare state policies in general, malaria epidemics of the scale of the one in 1934–1935 did not recur in Sri Lanka.

## The tsunami of 26 December 2004

The tsunami of December 2004 was certainly the biggest natural disaster to have occurred in Sri Lanka during the post-Independence era. In some respects, the tsunami had a much more dramatic impact on Sri Lanka society. In the same way as the worldwide economic depression of the 1930s set the stage for the malaria epidemic, the tsunami affected Sri Lanka at a time when the country was barely emerging from two decades of civil war. It was more of a globalized event than the malaria epidemic described above, with several countries simultaneously affected by a common disaster that evoked a global response.

To some extent, the victims of the malaria epidemic were taken by surprise as the epidemic broke out in wet and intermediate zones outside the malaria endemic region. The term 'malaria' itself was not widely known by many of the affected people at the time. On the other hand, the tsunami was a total surprise and a totally new experience to the affected people, having no prior knowledge whatsoever of the nature of the disaster. The tsunami waters devastated more than 70% of Sri Lanka's coastline and in their wake left more than 35,000 dead, 800,000 people homeless, 78,000 homes ravaged and uninhabitable, 200 schools, 72 hospitals, 84 of 242 hotels damaged, and 1615 km roads and 155 km of railway impassable.[3] Five of the nine provinces in Sri Lanka were affected. A total of 1162 villages and towns in coastal areas in Sri Lanka were directly affected by the tsunami. The damage to property and infrastructure was estimated to be USD 1 billion (ADB 2005).

Most importantly, the tsunami affected all three larger ethnic groups in the country, namely Sinhalese, Tamils and Muslims, initially creating a sense of common tragedy in an ethnically polarized nation. However, in terms of death and displacement, the worst affected were Muslim and Tamil minorities inhabiting the densely populated east coast, who had also been affected by decades of armed conflict and escalating ethnic tension. In contrast to the malaria epidemic of the 1930s, which broke out in impoverished Sinhala villages in western and central regions of the country, the tsunami mainly affected socially and politically marginalized Muslim and Tamil settlements in the eastern coastal belt and some of the mainstream Sinhala settlements in the south-western coastal belt (Shanmugaratnam 2005; Vaux et al. 2005; McGilvray 2006). While this created opportunities for interethnic solidarity and establishing bridging social capital among ethnic groups in the humanitarian response, such opportunities were not captured by the key players, namely the state, the LTTE and civil society organizations, in a way that creatively responded to the disaster and social and political realities in the country.[4]

### *The tsunami's impact*

As a natural disaster, the impact of the tsunami was far more intense and much more rapid compared to the malaria epidemic. Its full impact was felt in the affected areas within a very short time and wiping out lives, houses, livelihoods, and physical infrastructure of all kinds in a totally different way from the long and incremental human suffering experienced during an epidemic. Large-scale displacement of affected people and the consequent dislocation together with loss of family members were other distinctive attributes of the tsunami disaster. In war-affected north-eastern Sri Lanka, the tsunami simply produced a new wave of displacement and new misery, adding to people's plight of already entrenched social, economic and political instability, and intensifying conflict dynamics and the struggle for resources such as land, water and living space.

The vulnerability profile of the tsunami victims was in some ways similar to the victims of the epidemic. While marginal farmers and the landless in interior Sri Lanka were the primary victims of the epidemic, given the coastal nature of the tsunami disaster, fishermen were the worst affected, in terms of both loss of lives and property, including fishing gear, followed by wayside shop owners, hoteliers and workers in coastal establishments, including hotels, shops and salterns. The fishermen were most vulnerable not only because they lived so close to the sea, but also because their housing was such that it offered no resistance whatsoever to the violent tsunami waves. According to an estimate by the Department of Fisheries, almost 15% of all tsunami deaths in Sri Lanka were either of fishermen or their family members.[5] The total direct damage to the fisheries sector was estimated to be USD 335 million.[6] In north-east Sri Lanka, the insecurity created by the struggle between the government and the LTTE for naval supremacy had added to the vulnerability of fishermen long before the tsunami hit them. There is no accurate information about the age and gender distribution of deaths caused by the tsunami, but it is generally understood that women, children and the elderly were the greatest casualties. According to one estimate, more than two-thirds of the victims were women or children under the age of six years (de Silva 2005). Infants were particularly vulnerable in view of the inability of adults to save their lives by escaping the reach of the tsunami waves and in many instances women lost their lives in their efforts to save the lives of children (Harendra de Silva, personal communication 2005). Unlike the ethnically homogenous spread of the malaria epidemic, the victims were ethnically diverse with differential linkages with the state, as it had evolved in the post-independence era. Almost 46% of all tsunami deaths were in the two adjoining eastern coastal districts of Ampara and Batticaloa, with a predominance of Muslim and Tamil ethnic minorities, while Hambantota and Galle Districts in the Southern Province, with predominantly Sinhala populations, accounted for 28% of tsunami deaths (Thywissen & Renaud 2006).

One important feature of the tsunami impact in Sri Lanka is that it did not lead to any serious health problems, including epidemics. This speaks to the quality and accessibility of health services throughout the country, with the possible exception of LTTE-controlled areas. Even though the tsunami destroyed many health facilities along the coast, the government and international humanitarian agencies such as ICRC (International Committee of the Red Cross), Merlyn and Oxfam provided emergency health services where necessary. The comprehensive system of health services in Sri Lanka, in turn, can be largely attributed to the post 1934–1935 developments in health and welfare facilities in the country.

## Responses to the tsunami

In view of the extent of physical damage and large-scale displacement of people and also the suddenness with which the catastrophe occurred, the tsunami was immediately recognized as a serious humanitarian disaster. The instant worldwide telecast of news by leading electronic media, including CNN and the BBC, an aspect of globalization that was not operative in the 1930s, facilitated a global humanitarian response of an unprecedented nature.

The tsunami in Sri Lanka evoked local, national and international responses within a short period of time. Local communities, often led by religious institutions, came to the rescue of victims and organized relief activities instantly. Various voluntary groups collected donations of food, clothing and medicines from the public at large and supplied them to affected communities without consideration of their ethnic, religious or caste identities. As noted by one observer:

> One of the positive things we observed was the sudden shattering of barriers of ethnicity, religion, class, friend or foe, where human beings came together in support of each other, affected or otherwise, local or international, through pure emotions of humanity. People in affected areas, often with children, parents and/or friends dead and missing or their houses and belongings 'gone', came to each other's help. How they shared the food, looked after each other's children, elderly or belongings and comforted each other in spite of all the negative effects of the disaster, is something to be proud of. (de Silva 2005, 1)

This community mobilization driven by mass sympathy for the victims was similar to the response to the 1934–1935 epidemic. However, as organized humanitarian response from the state, NGOs and international community gathered momentum and aid flows became established, local community initiatives gradually fizzled out. In some instances, the sympathy of the non-affected gradually turned into resentment towards the beneficiaries of aid who received all the attention from external agencies.

Instant reporting of the disaster through global electronic media particularly led to the quick mobilization of an international humanitarian response from all over the world. Described as the biggest global media event subsequent to '9/11' (11 September 2001 attacks by al-Qaeda on the United States), media depiction in Asia focused heavily on child victims of the disaster. For instance, 'baby 81' who survived the tsunami waves and was claimed by many parents with missing children became something of an 'icon of tsunami' in eastern Sri Lanka (Jabry 2005). Largely as an outcome of the global media attention, 'one of the largest relief and rehabilitation operations ever launched by humanitarian organizations around the world' became implemented in Sri Lanka and other affected countries in Asia (Rawal et al. 2005). This involved UN agencies, international and local NGOs, various government agencies, and private sector organizations of various kinds. A directory prepared by UN Humanitarian Information Centre for Sri Lanka (HIC) in January 2005 lists a total of 890 agencies involved in post-tsunami operations in Sri Lanka (summarized in Table 2).

Many of the local and international NGOs already well established in Sri Lanka rapidly expanded their relief portfolios in response to tsunami. In agencies such as CARE International and World Vision Lanka the number of staff and scale of operations rapidly expanded by more than 100% in some instances (CARE International 2005; World Vision 2005). A large number of international NGOs arrived in Sri Lanka in response to the tsunami, often

*Table 2.* Number of humanitarian agencies involved in tsunami work by type as of January 2005

| Type of organization | Number | No. of child-focused organizations |
|---|---|---|
| Government | 86 | 3 |
| International donors | 18 | 0 |
| International NGOs | 353 | 15 |
| National NGOs | 367 | 21 |
| UN | 36 | 1 |
| Private sector | 30 | 0 |
| Total | 890 | 40 |

*Source:* Humanitarian Information Centre directory, 2005.

leading to intense competition among agencies for territory, beneficiaries and recognition by other actors (de Silva 2005; McGilvray 2006). In order to coordinate the activities of donors, NGOs and relevant state agencies, the Centre for National Operations (CNO) was established in the Presidential Secretariat with wide powers for coordinating the relief efforts (Centre for National Operations 2005). This was later replaced by the Task Force for Relief (TAFOR) and the Task Force for Rebuilding the Nation (TAFREN), in charge of relief work including temporary shelter and housing and infrastructural development respectively. Finally, a centralized state agency, the Reconstruction and Development Agency (RADA), staffed by political appointees, took over the overall functions of coordinating reconstruction. This, in turn, marked the transition from relief to recovery and from reconstruction to development which called for more sustained efforts and more systematic implementation mechanisms. In 2006, a Ministry of Disaster Management was established for coordinating disaster monitoring and disaster prevention activities. Similarly, the United Nations Office for the Coordination of Humanitarian Affairs (OCHA) was established for coordinating and streamlining the activities of UN agencies involved in post-tsunami operations (UN 2005). Among the international and local NGOs too, several initiatives for increased coordination of activities have evolved in recent years, with OCHA and joint evaluations of tsunami response as examples of such initiatives.

## *Politics of disaster management*

In keeping with the central role played by the state in previous disasters and the long experience of the Sri Lanka welfare state, the state undertook to lead the relief and rehabilitation efforts in the tsunami response. A new administrative machinery directly under the control of the President of Sri Lanka was created for coordinating relief, recovery and rehabilitation operations. As was also resorted to during the malaria epidemic, new legal measures were introduced, on this occasion through the Tsunami Act of 2005, to ensure the rights and protection of various parties, including the children who lost their parents during tsunami, and also the Disaster Management Act of 2006. The state introduced a ban on house rebuilding within a specified buffer zone close to the shore, and while this was done with good intentions it made it difficult for the victims of the disaster to secure permanent housing. The model of a paternalistic welfare state continues to be projected by the state, but the state's legitimacy has been vastly eroded over the past several decades due to a number of internal and external pressures, including the war. Korf (2005; 2007), for instance, has described the contemporary Sri Lankan state as a 'failed state'.

The capacity of the state to effectively carry out its relief and recovery operations, however, was limited, certainly by comparison to the 1934–1935 epidemic due to perceived partiality of the state towards the Sinhalese, the majority ethnic group, the greater damage being in the East Coast of Sri Lanka where ethnic tension was running high and where some of the territory was controlled by the LTTE. The Post Tsunami Operations and Management Scheme (PTOMS), proposed as a mechanism whereby the state and the LTTE were expected to jointly implement the relief and recovery operations in respective areas, could not be implemented due to, on the one hand, the keenness on the part of the LTTE to secure direct funding from external donors for relief operations in ways that would legitimize its power in the north-east and, on the other, the government's motive to control resource flows in ways that reinforced and reinvigorated the power of the Centre. Unlike the 1934–1935 epidemic, when the state had undisputed authority over the affected areas and was increasingly driven by representatives of the affected people, the power and authority of the state in relief and recovery operations in northern and, to a lesser extent, eastern Sri Lanka was strongly contested by the LTTE, which considered these regions to be parts of the Tamil homeland (*Tamileelam*) and therefore under the jurisdiction of the LTTE, as specified by the MOU signed with the Government of Sri Lanka in 2002. Thus, in contrast to responses to the 1934–1835 epidemic which paved the way for the development of a welfare state responsive to the needs of wider masses, the tsunami response revealed the fractured nature of Sri Lanka society and polity. Moreover, intense competition for control of aid flows introduced a new dimension to the ethnic conflict in Sri Lanka and may have had the effect of intensifying the conflict in contrast to the positive impact of the tsunami in facilitating a peace process in Aceh Province in Indonesia (Sirisena 2005).

By comparison to the malaria epidemic of 1934–35, the humanitarian response to the tsunami was much more globalized and was guided by international standards such as Sphere principles and the ICRC code of ethics, but this did not necessarily mean that the response was better organized or even more effective (Vaux et al. 2005). As many as 500 international NGOs arrived in Sri Lanka in response to the disaster. In addition, the INGOs (international non-government organizations) already active in Sri Lanka, such as CARE International, PLAN International and World Vision Lanka, quickly modified their aid portfolios to assist the survivors. A significant part of the aid flow was used to support the super salaries of expatriate staff, their luxury vehicles and their exceedingly high living costs. de Silva (2005), for instance, notes that the daily allowance paid to some of the international experts involved in tsunami humanitarian efforts in Sri Lanka often exceeded the monthly salary of persons with similar qualifications in Sri Lanka. Some of the INGOs went into partnership with local NGOs in order to implement their activities but others opted to implement their programmes directly in territories totally unfamiliar to them. For instance, a number of new agencies that arrived in Sri Lanka in the aftermath of tsunami conducted some psychosocial interventions in the affected areas without checking their potential adverse effects (A. Galappatti, personal communication 2006). Kusala Wettasinghe (2005), who had conducted consultations with tsunami-affected children regarding their experiences during the tsunami response, found that the same children had been used in successive psychosocial support programmes by different agencies without considering what impact these diverse interventions would have on these children.

The INGOs and NGOs distributed themselves in different districts and different sectors (e.g. housing, water and sanitation, health, psycho-social support, and livelihood development), but there was limited coordination among different players and there was frequent competition for territory. Some of the existing programmes heavily suffered as the INGOs pulled out resources and personnel from such programmes to more lucrative tsunami relief and recovery operations with enhanced funding (CARE International 2005; Save the Children 2005; World Vision 2005). There was considerable resentment within the organization when salaries paid to tsunami workers were many times higher compared to their counterparts in other operations. Some INGOs recruited top-level expatriate staff from European countries where the agency headquarters are located, mid-level expatriate staff from India, and low-level local staff from Sri Lanka, reflecting a hierarchical pattern of organization that is well entrenched in many aid organizations in Sri Lanka. Some of the local NGOs were pushed to the background as they could not match the resources controlled by big players in the scene. Also, some local NGOs were unwilling to partner these big players under the terms and conditions dictated by the latter. As already noted, the local humanitarian response to the disaster motivated by sympathy towards the victims, voluntarism and humanitarian concern irrespective of caste, creed or ethnicity quickly disappeared in the wake of international humanitarian response driven by aid flows in US dollars and high-tech interventions, sometimes with limited sensitivity towards local cultures. For instance, Islamic groups as well as LTTE imposed a ban on some of the INGOs operating in Eastern Sri Lanka, claiming that they disregarded local culture particularly in assigning duties and tasks to local female staff (LankaNewspapers.com 2006).

The centralized state structures established for coordinating the relief, recovery and rehabilitation operations secured pledges from various donor agencies, INGOs and local NGOs for specific activities. The ruling political elite often used tsunami relief to strengthen and extend their existing political bases. In some instances, this had the effect of leakage of tsunami benefits to those unaffected by the disaster, and at the same time diverting such benefits away from the real victims of the disaster as will be shown later in this paper. The political structure of the country was such that aid flows to regions 'convulsed by the dual traumas of civil war and tsunami', i.e. Northern and Eastern Provinces, were hampered in many ways (McGilvray 2006). Unlike the malaria epidemic of 1930s that paved the way for the development of a welfare state with universal coverage of free health care and free education, the tsunami response clearly reinforced the existing political and ethnic prejudices of the state.

Housing received highest priority in the recovery and reconstruction efforts led by the state and its NGO partners. The total number of houses fully and partially damaged was estimated to be 118,000. Some displaced people opted to live with their relatives outside the damage zone. Others moved to temporary forms of shelter and subsequently to transitional shelters provided by the NGOs and INGOs. Two methods of support for housing construction were employed. For those outside the buffer zone, a fixed grant was given by the state to help them rebuild their own houses, depending on whether their house had been fully damaged or partially damaged (IPS 2006). For those displaced from within the stipulated buffer zone, housing projects were to be constructed in alternative venues with the support of the donors and NGOs. The idea of housing projects appealed to NGOs as it enhanced their visibility and ability to demonstrate 'concrete' results, particularly to headquarters' staff who came to visit on monitoring missions. Such housing projects were even more appealing to local and national politicians, as they enhanced their political base and patronage resources. Where the venues of such housing projects were identified early, prominent signboards were erected by the agencies, sometimes with pictures of key politicians, long before the projects materialized. In some instances, the projects did not progress beyond the construction of the signboards. The housing options that the victims themselves preferred were not assessed systematically, even though some of the NGOs that were committed to participatory development consulted the beneficiaries regarding the housing design they wanted and involved them in monitoring the process. Sometimes the NGOs received the beneficiary lists from relevant government agencies only after the houses had been built. In many instances, the houses were built by contractors employed by the NGOs without any participation by the intended beneficiaries. It is worth noting that 'owner-driven' housing development seems to have been more effective, both in terms of the rate of successful completion and the quality of housing achieved (IPS 2006).

In October 2007, RADA estimated that new houses to replace more than 80% of the houses destroyed by tsunami would be built by the end of the year (Perera 2007). This, however, appears to have been an unrealistic expectation given the actual results at the ground level. While the reconstruction efforts in the south seem to have proceeded on schedule, in the war-affected north and east, including some of the areas worst affected by the tsunami, progress has been much slower and beset with many obstacles. In a report in the series 'Voices from the field', dated October 2007, Action Aid noted that less than 12% of damaged houses had been rebuilt or reconstructed, as compared to 26% in the east and 86% in the south of the country (Perera 2007). A more systematic sample survey of affected households in the east and south conducted by IPS in July 2006 indicated somewhat similar results (IPS 2006). There is also considerable evidence of significant leakage of tsunami funds in the north and south to political followers of the ruling regimes. This, in turn, raises the question of why the wave of tsunami aid did not produce satisfactory results four years into recovery and reconstruction.

Instead of empowering the affected people, the individual-centred (e.g. direct benefits to affected individuals and households), technocratic (housing, sanitation, health services, education) and uncoordinated tsunami response in Sri Lanka had the effect of promoting dependency among the people and eroding the capacity of communities to bounce back and play a prominent role in their own recovery after the disaster.

## *'The third wave'*

The tsunami response in Sri Lanka is by no means unusual, as the global attention on disaster victims in the modern world tends to be short-lived and highly media driven. The Sri Lankan case points to the dynamic interaction among local, national and global processes that intervene in the process of recovery and reconstruction in disaster response. Focusing on the housing issue, several interconnected and interwoven factors contributed to limited progress in house building in the tsunami response in Sri Lanka. First, there occurred a flight of INGOs after an initial spurt of activity centred on relief and recovery operations. In this article, the term 'tsunami third wave' is used figuratively to refer to this rapid withdrawal of many INGOs that originally came to provide humanitarian assistance in the aftermath of the 'second tsunami wave', and the resulting disruption in recovery and rebuilding processes. This was partly due to the local situation, such as resumption of fighting, but also due to movement of funds and personnel from one global disaster to another in a world full of disasters. For instance, the earthquake in Pakistan encouraged some INGOs to divert attention to the new flashpoint before they had completed their mission in Sri Lanka. TAFREN, and later also RADA, pointed out that many tsunami housing projects had to be abandoned as the funds pledged by some donors were not forthcoming. In a statement to the press on 23 December 2007, RADA mentioned that of the total pledge of USD 3.1 billion, only USD 1.7 billion had been actually committed or released by the donors by that date. On the other hand, donors found fault with the state for inappropriate polices (for instance, a ban on rebuilding houses within a buffer zone and subsequent alteration of the buffer zone on a number of occasions, upsetting existing plans for housing construction) and not fulfilling its obligations particularly in the supply of land for housing construction, as described below.

Second, one of the controversial policies that the government introduced was the declaration of a coastal buffer zone where housing construction, including rebuilding of houses damaged by tsunami, was not permitted as a preventive measure against any future tsunamis. Some critics argued that this policy was motivated by a neo-liberal push to drive away victims to make room for the hotel industry, particularly on the east coast (People's Planning Commission 2007). The buffer zone meant that the state had to provide alternative land for the construction of housing for many displaced people. The implications of this policy for housing the victims were not seriously considered, nor was there any consultation with the affected people before this policy was formulated (Shanmugaratnam 2005). This appears to have been a decision taken by a few very senior bureaucrats without consulting affected people and any social scientists with relevant knowledge (McGilvray 2006). Finding alternative land and locations acceptable to tsunami Internally Displaced Persons (IDPs) and the potential host communities in the surrounding areas posed many challenges. The relocation of IDPs presented the most serious challenges on the east coast, where the buffer zone was twice the width that it was in remaining parts of Sri Lanka, presumably due to the greater vulnerability of the east coast for tsunamis emanating from earthquake-prone areas in the Indonesian archipelago. The problems on the east coast were due to the narrow stretch of land suitable for building between the sea and lagoons and other low lying areas, including vast stretches of paddy fields, and also the possible implications of any large-scale relocation of IDPs of minority ethnic status to areas further inland with a Sinhala predominance, in view of possible ethnic tensions it may have given rise to (Shanmugaratnam 2005). Even in the densely populated south-western coastal belt, similar problems arose due to scarcity of land. Tension characterized by a high degree of aggressiveness and assertiveness erupted between relocated persons mostly from coastal fishing communities and surrounding host communities exhibiting marked differences in lifestyle, dialect, dress code, and food habits.. The NGOs, however, began to work on housing projects where land was available and initial difficulties were overcome. In the meantime, by 2006 the state was compelled to change its policy on the coastal buffer zone, reducing it substantially due to the operational difficulties that the original declaration posed for establishing permanent housing for tsunami IDPs. Once again, a policy shift was introduced without seriously considering its implications for the housing projects already being designed or implemented. This led to some confusion regarding who was to be entitled to housing in the projects already being developed, causing further delays in implementation.

Third, resumed fighting between the government and the LTTE after December 2005 seriously affected housing construction in the north and east. On the one hand, this has made it difficult to send supplies and workers to construction sites in the conflict-affected areas. On the other hand, fresh outbreaks of violence have resulted in a new wave of displacement, estimated to have involved 300,000 people during the past two years, and the INGOs working in the affected areas have diverted attention to the supply of relief to these new IDPs, brought about by the conflict rather than by the tsunami. For instance, the International Federation of Red Cross and Red Crescent Societies, which undertook to build 26,000 housing units for tsunami victims in Mulaitiv (in the Northern Province) and other frontline areas, have frozen such projects until such time as the security situation improves and have undertaken relief operations for the new wave of IDPs resulting from aerial attacks in relevant areas. Similarly, World Vision Lanka put a hold on their housing projects in frontline areas (Senaratne 2007). This indicates that before recovery from the previous disaster is over, a new disaster has set in, complicating the recovery process. For many people this means that the temporary housing strategies they have resorted to – such as the transitional shelters in which they live or shared accommodation with family members – will turn out to be a long-term strategy with associated risks and uncertainties, such as rapid deterioration of housing conditions and increased tensions with host families.

Fourth, the flow of funds for tsunami-affected people has been seriously distorted by political interference in the allocation of funds at various levels. According to RADA, the Ampara District, in the Eastern Province, which suffered 24% of total housing damage, received 14% or USD 58 million of total pledges as against Hambantota, which sustained much less house damage yet received USD 45 million, which has been estimated to be almost five times its actual requirement (Perera 2007). In the massive Siribopura housing scheme in Hambantota, the political base of the current President of Sri Lanka, more than 30% of housing recipients are estimated to be political nominees who have not been affected by the tsunami at all (personal communication, local NGO project leader 2006). Some households managed to secure more than one house in the same housing scheme or sometimes in different housing schemes. This indicates serious targeting errors in the distribution of funds, thus affecting the overall effectiveness of the exercise. Where newly built houses have been provided to victims in new housing schemes, the beneficiaries have reported poor quality of housing, water and sanitation services, vulnerability to environmental hazards such as floods, and also the poor accessibility of new places of residence compared to their previous places of residence, which were usually on the beachfront (IPS 2006; Senaratne 2007). Thus, in some ways, both the beneficiaries and non-beneficiaries of houses seem to have grievances over their housing situation.

Fifth, organized civil society response from within Sri Lanka has been increasingly organized along ethnic and religious lines in a way that was uncommon in the past. For instance, the Darmavijaya Foundation, Tamil Rehabilitation Organization, and Islamic Relief have been formed to undertake relief work predominantly, if not exclusively, within their respective ethnic groups even though the disaster affected all three ethnic groups and the initial spontaneous humanitarian response to the disaster from within the country was very much broad based. For example, Muslim and Tamil victims of the disaster were both accommodated and assisted in local Hindu shrines in the immediate aftermath of the disaster. In the same way that political parties have become ethnically polarized during the past few decades, also some of the civil society organizations have developed along ethnic lines, making it difficult for them to work collectively for advancing the common interests of tsunami victims, vis-à-vis the state and the donors. On the other hand, international donors have often established partnerships with and utilized these different civil society entities in order to expand their operations within each of the affected ethnic communities. With the gradual departure of many international donors under the third wave, such ethnic-specific arms of civil society have once again become insular and unable to forge bridging social capital necessary for the development of a more effective civil society.

Finally, tsunami relief and recovery operations by numerous agencies without proper coordination between them or plans for cost recovery of some kind have a led to a high sense of dependency among the beneficiaries. For those with the 'right' connections to the Government Organizations (GOs) and NGOs, assistance from outside has become a disincentive for developing any sustainable livelihoods and even moving away from transitional shelters, once again interfering with their recovery process and ability to stand on their own feet. This recipient mentality must be seen as a yet another legacy of the third wave in which high expectations were produced but not matched by the reality. As has been pointed out in many evaluations of tsunami projects (Rawal et al. 2005),[7] the nature of aid flow itself has been partly responsible for this situation. In a sense, those who heavily depended on INGOs and NGOs for their survival are those who are most disempowered by the third wave, with no more advantages to be derived from their tsunami victimhood, often utilized by them as a blessing in disguise.

## Conclusions

This paper has examined the social and political consequences of two disasters that struck Sri Lanka 70 years apart. To return to the analytical question posed at the beginning, what accounts for the relative success of the post-epidemic developments in Sri Lanka during the 1930s and 1940s and the seemingly problematic outcomes of the tsunami response during the past four years? While the latter observation itself may be questioned and debated, it does not speak well for the global aid industry, well funded and closely guided by international standards and conventions. At the same time, it must be stressed that the political crisis in Sri Lanka and the failure of the ruling governments to satisfactorily address the legitimate grievances of ethnic minorities have made the tsunami response much more problematic and even counterproductive. On the whole, the post-epidemic response galvanized the Sri Lankan welfare state and civil society in a way not at all evident in the more glamorous but aid-driven tsunami response.

Unlike the 1934–1935 malaria epidemic that unfolded over an eight-month period and evoking several lasting responses from the emerging Sri Lanka state and civil society, the massive aid flows associated with the tsunami disaster led to intense competition between institutions involved in relief and recovery, further polarization of conflicting parties, and growth of dependency among the people. The third wave swept away many of the agencies that came after the tsunami and the unrealistic promises and hopes that came with them. Even though the government and the other political actors sought to take advantage of the impacts of the tsunami with their narrow political agendas, this did not lead to any kind of political mobilization of the affected people in ways that assured their rights and enhanced their participation in the decision making processes in state and non-state agencies. Even though the language of rights (e.g. children's rights, women's rights) was often used in donor policies, NGO-led programmes and government legislations (e.g. the Tsunami Act of 2005), it was not translated into practice in programmes essentially directed from the centre, but fragmented due to ethnic polarization and conflicting political loyalties and affiliations among affected people. What was initially identified as a common tragedy requiring joint action quickly gave way to competing interests articulated along ethnic, regional and political party lines. Interestingly, the tsunami did not produce an equivalent of the Suriya Mal Movement or leftist political parties that campaigned for the rights of disaster victims in and outside the legislature.

On the whole, in the globalized post-modern era characterized by instant worldwide publicity of disasters through global electronic media, disaster response is driven by technology, aid flows and transnational agencies with limited or no accountability to affected states, their peoples and civil society organizations. The tsunami's third wave is symptomatic of this global aid industry, in that it moves in fast where disaster strikes, provides emergency relief, and moves on to the next disaster site without necessarily completing the tasks undertaken. Most importantly, in terms of social and political processes it does not build on any existing processes or leave behind any viable entities that add to the resilience and coping mechanisms of the affected populations.

## Notes

1 An earlier version of this paper was read at an international conference on 'Social Science Perspectives relating to Post-tsunami Recovery in South Asia' held in Jawaharlal Nehru University, New Delhi, 14–15 January 2008.
2 It is not claimed here, however, that this disaster response was the only cause of the development of a welfare state in Sri Lanka. Some enlightened social policies initiated in the latter part of the British rule, as well as the importance of Ceylon as a testing ground for some of these policies in the British Empire, including application of tropical medicine, have contributed to these developments.
3 TAFREN (Task Force for Rebuilding the Nation) news release, July 2005.
4 This was in sharp contrast to the peace accord signed between the Government of Indonesia and Arce liberation movement in the aftermath of the tsunami.
5 http://www.fisheriesdept.gov.lk/tsunami/deaths.html (accessed August 2008)
6 http:/www.streaminitiative.org/pdf/Srilanka (accessed March 2005[0])
7 See also J. Markvosen, 'Assessment of psychosocial and educational needs in tsunami stricken areas in Sri Lanka. HNI-TPO, Colombo, 2005 (unpublished report).

## References


ADB (Asian Development Bank). 2005. *Asian Tsunami Impact: Preliminary Damage and Needs Assessment for Sri Lanka*. ADB, Manila.

Briercliffe, R. 1935. *Ceylon Malaria Epidemic, 1934-35: Report by the Director of Medical and Sanitary Services.* Paper 22 of 1935. Ceylon Government Press, Colombo, Sessional Manila.

CARE International. 2005. *Post-tsunami Community Development Program, the First One Hundred Days.* CARE International, Colombo.

Centre for National Operations. 2005. *Policy Framework for the Protection and Care of Children Affected by the Tsunami Disaster.* CNO, Colombo.

de Silva, H. 2005. Human and social tragedy in the aftermath of tsunami: Some thoughts. *Asian Tribune* June 2005, 5–9.

Gill, C.A. 1935. *Report of the Malaria Epidemic in Ceylon 1934/35*. Sessional Paper 23 of 1935. Government Printer, Colombo.

Government of Ceylon. 1935. *State Council of Ceylon Debates.* Government Printer, Colombo.

Harrison, G. 1978. *Mosquitoes, Malaria and Man: A History of Hostilities since 1880*. E.P. Dutton, New York.

IPS (Institute of Policy Studies). 2006. *Post-tsunami Reconstruction and Rehabilitation: Household Views on Progress and Process.* IPS, Colombo.

Jabry, A. 2005. *After the Cameras Have Gone: Children in Disasters.* Plan Limited, Woking.

Jayasuriya, L. 2000. *Welfarism and Politics in Sri Lanka: Experiences of a Third World Welfare State.* University of Western Australia, Perth.

Jones, M. 2000. The Ceylon Malaria epidemic of 1934-35: A case study in colonial medicine. *Social History of Medicine* 13:1, 87–109.

Korf, B. 2005. Sri Lanka: The tsunami after the tsunami. *International Development Planning Review* 27:3, i–vii.

Korf, B. 2007. Antinomies of generosity: Moral geographies and post-tsunami aid in Southeast Asia. *Geoforum* 38:2, 366–378.

LankaNewspapers.com. 2006. NGO porn scandal shocks Batti, Ampara. 23 April 2006. http://www.lankanewspapers.com/news/2006/4/6567.html (accessed October 2008).

Lerski, G.J. 1969. *Origins of Trotskysm in Ceylon: A Documentary History of the Lanka Samasamaja Party.* Hoover Institution, Stanford, CA.

McGilvray, D. 2006. Tsunami and civil war in Sri Lanka. *India Review* 5, 372–393.

Meyer, E. 1982. L'epidemie de malaria de 1934-35 a Sri Lanka: Fluctuations economiques et fluctuations climatiques. *Cultures et Development* 14:2-3, 183–226.

Newnham, H.E. 1935. *Report on Relief of Distress due to Sickness and Shortages of Food, September 1934 to December, 1935.* Sessional Paper of V of 1936. Ceylon Government Press, Colombo.

People's Planning Commission. 2007. *Land and Housing Rights in the Post-tsunami Context*. People's Planning Commission, Colombo.

Perera, A. 2007. *Sri Lanka: Not All Tsunami Reconstruction is Equal.* http://www.ipsnews.net (accessed December 2007).

Rawal, V., Fautin, C., Moore, J.-L., Kalonge, S., Walden, V.M. & Bhattacharjee, A. 2005. *Realtime Evaluation of Multi-agency Tsunami Response in Sri Lanka and India*. CARE International, Oxfam and World Vision International, London.

Save the Children . 2005. *Six Monthly Progress Report on Tsunami Response.* Save the Children, Colombo.

Sen, A. 1982. *Poverty and Famines: An Essay on Entitlements and Deprivation.* Clarendon Press, Oxford.

Senaratne, M. 2007. Tsunami billions still at sea: Thousands of victims stranded three years after the catastrophe. *The Sunday Times* 23 December 2007, 3. Colombo.

Seneviratne, H.L. 1999. *The Work of Kings: The New Buddhism in Sri Lanka.* Chicago University Press, Chicago.

Shanmugaratnam, N. 2005. Tsunami victims' perception of the buffer zone. *Polity* 2:4, 13–15.

Silva, K.T. 1994. Malaria eradication as a legacy of colonial discourse: The case of Sri Lanka. *Parassitologia* 36:1-2, 149–163.

Silva, K.T. 1997. Public health' for whose benefit? Multiple discourses on malaria in Sri Lanka. *Medical Anthropology* 17:3, 195–214.

Sirisena, M. 2005. Old habits die hard: Nationhood in the aftermath of tsunami. *Polity* 2, 11–12.

Sybil, X. 1963. Birth of the Left Movement: A May Day reminiscence. *Forward: Progressive Weekly* 10, 4–10.

Thywissen, K. & Renaud, F. 2006. *Devastation and Reconstruction: The Situation in Sri Lanka After the 2004 Tsunami: Observations from 3 UNU-EHS missions to Sri Lanka in February 2005, July 2005 and January 2006.* http://www.ehs.unu.edu/article:209?menu=73 (accessed January 2008).

UN . 2005. *UN Strategy Paper: Coordination of the International Response in Tsunami-affected Areas in Sri Lanka*. OCHA, Colombo.

Vaux, T., Bhatt, M., Bhattacharjee, A., Lipner, M., McCluskey, J., Naik, A. & Stevenson, F. 2005. *Independent Evaluation of the DEC Tsunami Crisis Response*. Valid International, London.

Wettasinghe, K. 2005. *Report on a Consultation with Tsunami-affected Children in Sri Lanka*. Plan International, Colombo.

Wigglesworth, V.B. 1936. Malaria in Ceylon. *Asiatic Review* 32, 611–619.

World Vision. 2005. *Sri Lanka Tsunami Response Sector Report*. World Vision Lanka, Colombo.

# Is 'build back better' a response to vulnerability? Analysis of the post-tsunami humanitarian interventions in Sri Lanka

SARAH KHASALAMWA

Post-crisis situations represent a high level of very demanding and complex needs. The longer term goal of recovery should address vulnerabilities to future disasters that are likely to occur with increasing frequency and often in the same places. Post-crisis recovery should not be merely a return to the status quo ante but an attainment of a 'new normalcy'. 'Build back better', as advocated by President Bill Jefferson Clinton after the 2004 tsunami, underlines the need for quality in the recovery efforts. In the paper, recovery is interpreted as addressing the root causes of vulnerability and the pre-existing processes and conditions from which they stem. The paper explores how the narrative 'build back better' is perceived by different actors and how these perceptions are manifested in their operational priorities and programmes. An attempt is also made to examine the outcomes. This is discussed in relation to the post-tsunami recovery in Sri Lanka, specifically examining the interventions of selected humanitarian organizations visited during the research. The paper argues that despite the engaging mantra 'build back better', the tsunami response has not lived up to expectations, nor has it significantly altered the existing structural vulnerabilities.

*Sarah Khasalamwa, Department of Geography, Norwegian University of Science and Technology (NTNU)*

## Introduction

The aftermath of the earthquake and tsunami disaster on 26 December 2004 saw an unprecedented humanitarian response, considered one of the largest and most generous in the history of the humanitarian efforts. Approximately USD 13.5 billion was raised immediately (Flint & Goyder 2006). There have been significant achievements in recovery and reconstruction in the affected countries ranging from provision of shelter to restoration of livelihoods. Increasingly, post-crisis recovery is viewed as opportunity to vitalize or revitalize the development process through regeneration of the local economy, upgrading livelihoods and living conditions for the affected people. The post-tsunami recovery provides a learning opportunity for whether to validate the assumption. The major debate is about how to reduce vulnerabilities and enhance capacities at different levels to cope with future crises. This involves building sustainable communities and not recreating or exacerbating vulnerabilities (Wisner et al. 2004). The goal of recovery is not merely a return to the status quo ante but an attainment of a 'new normalcy'. With regard to the tsunami recovery efforts, the phrase 'build back better' was adopted to depict a more comprehensive approach to the recovery process. The UN Special envoy for Tsunami Recovery, William Clinton has stressed the need for a better development path and empowering communities to cope better with future disasters.

This paper examines the humanitarian interventions and particularly how the interventions contribute to the reduction of vulnerability. Using evidence from a research trip to Sri Lanka in August 2007, I explore the work of selected humanitarian agencies and International Non-Governmental Organizations (INGOs) as a way of understanding perceptions regarding 'build back better' and how these have been manifested in the different programmes and priorities. The organizations visited were the International Federation of Red Cross and Red Crescent Societies (IFRC) and Oxfam Great Britain (Oxfam GB). I visited some projects that were being implemented by the selected organizations in Hambantota District (Southern Province) and Ampara District (Eastern Province) (Fig. 1). This was basically a qualitative study. I conducted interviews with the humanitarian agencies' officials in Colombo and their field staff as well as the project beneficiaries. With regard to the beneficiaries, I visited cases selected by the organizations. This was the most practical approach to identifying households and individuals who were benefiting from the organizations' programmes. Furthermore, there are hundreds of Non-Governmental Organizations operating in the selected areas and it was problematic to attribute the outcomes without being specific. In Hambantota it was relatively easy to speak to the project stakeholders, although an attempt to meet beneficiaries in the housing projects on my own was met with hostility. I tried to select randomly some of respondents who refused to talk to me. When my translator (a native) tried to take a general photograph of the area one man emerged and had a bitter exchange with him. His argument was that they were tired of researchers and tourists who were apparently making fun out of their misery. Fearing repercussions, I did not persist and decided to leave. Meanwhile in the east the situation was tense when I conducted my fieldwork. When I visited Bakkiella village, located *c.*25 km north of Ampara town, close to Batticaloa, it was possible to hear the exchange of gunfire. There was restriction on visitors and for this reason it was important to visit under the umbrella of an

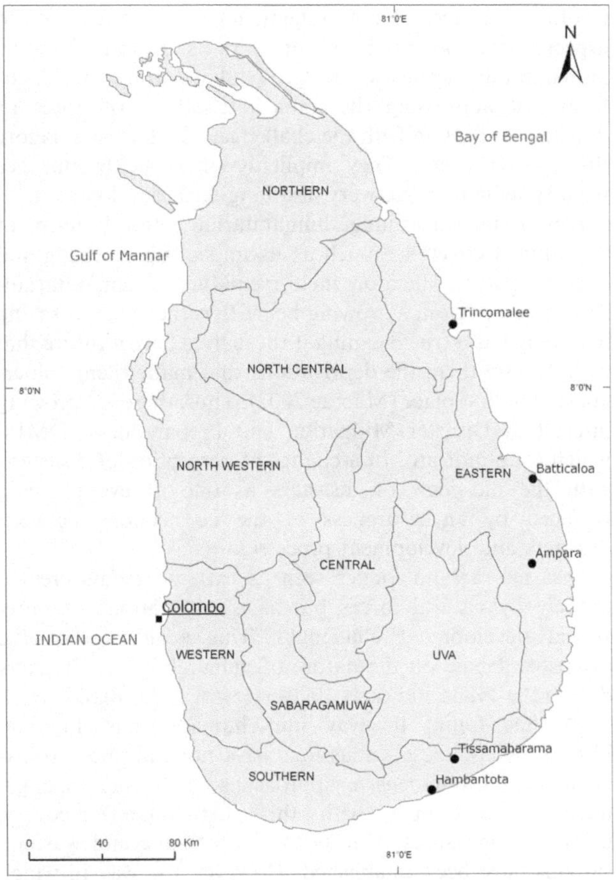


*Fig. 1.* Sri Lanka: location of provinces and selected places mentioned in the text

organization. No photographs were allowed to be taken. In this area, I encountered some respondents who were very emotional about their situation. In such cases, I was very selective and sensitive about the questions I asked. I also relied on direct observation where possible. In some of the areas I visited in Ampara there was little taking place in terms of actual interventions, especially in the recently cleared areas, and only needs assessments had been done. The IFRC attributes the lag to the insecurity that has prevailed in the area. Nevertheless, this was an opportunity for first-hand experience of the existent challenges. One major drawback with the methodology was the limited choice regarding the respondents and bias introduced by the presence of organization officials during some interviews. In areas where needs assessments were in progress the officials appreciated the information generated. It was a learning opportunity for both parties. However, I must state that the organizations were very open to learning about the challenges of their beneficiaries, especially in implementation of the programmes. This was reflected in the selection, where there was representation of both the successful cases and the challenged ones. However, given the time constraint and security concerns the data have been supplemented by an extensive review of secondary information provided by the organizations. This has also been backed with information from the Internet in the form of annual reports, project evaluations, news articles, and other relevant publications.

## Theoretical perspectives

### *Understanding vulnerability*

Conventionally, vulnerability is viewed as a set of attributes that condition outcomes as *susceptibility* to impact (Lewis 1999). Wisner et al. (2004, 11) define vulnerability as the 'characteristics of a person or group and their situation that influence their capacity to anticipate, cope with, resist and recover from the impact of a natural hazard (an extreme natural event or process)'. Vulnerability is viewed in terms of variations in exposure to hazards as well as variations in people's capacity to cope with hazards. It is a concept that combines exposure to a threat with susceptibility or sensitivity to the adverse consequences. It is argued that disaster vulnerability is socially constructed – a result of the social and economic circumstances of everyday living (Morrow 1999). Also, the nature and severity of the disaster influences vulnerability. The nature of disaster determines vulnerability. For example, the tsunami as a natural disaster was vast both in severity and coverage. Usually, the impact of natural disasters is vast (e.g. the recent earthquake in China and the Burma cyclone both caused massive destruction) compared to that of disasters which are principally man-made, and which are often confined to specific areas. Man-made disasters can be equally devastating, such as the bombing of Nagasaki and Hiroshima, and the Bhopal chemical spillage that caused massive destruction. However, in other man-made disasters, such as wars, the impact and vulnerability tend to develop gradually. Nevertheless, irrespective of the nature of the disaster, those who are most affected are those who are already vulnerable. Wisner et al. (2004) offer a more holistic view of disasters and their impacts, and consider both natural as well as social causes of disaster and the interaction between the two. They acknowledge the fact that natural events trigger disasters but also emphasize the various ways in which social systems operate to generate disasters by making people vulnerable. Local disaster vulnerability is increasingly recognized to be the pre-existing patterns of community settlement and development (Anderson 1994; Varley 1994).

The impact of a natural event on any given community is not random but determined by everyday patterns of social interaction and organization, particularly the socio-economic processes which determine the degree and distribution of access to resources (Blaikie et al. 1994; Bolin & Stanford 1998). The effect on any particular household is determined by a complex set of interacting conditions such as the geography and location, some with dwelling place characteristics and others with social and economic characteristics of the people living there (Hewitt 1983; Quarantelli 1987; 1995). This is clearly evident in the areas affected by the tsunami of 2004, where the geographical location was critical in determining who suffered the most impact. Understanding the impact of disaster as mainly a result of pre-existing conditions can be a basis for designing interventions that will not only address the stress resulting from the impact but also transform the affected community in the future. In this paper, 'build back better' has been conceptualized in relation to vulnerability because the proponents argue that the post-tsunami recovery efforts should not only address the needs created by the tsunami disaster but also the pre-existing challenges as a way of increasing resilience and enabling communities to cope better with future disasters.

Post-disaster recovery interventions must address factors that influence the vulnerability of individuals, including their socio-economic status, gender, age, health status, ethnicity, and political or religious affiliation, and also the geographical locations, infrastructure and social networks. Blaikie et al. (1994) also emphasize the need to recognize the differential effects of natural disasters on neighbourhoods and even households. The geography of vulnerability is therefore not the same as the geography of hazard. It also follows that the differential impact requires customized interventions rather than standardized response. In the case of the tsunami of 2004 there was a tendency for standardized responses that did not specifically address the needs of the affected communities, as will be discussed later in the paper. Morrow (1999) stresses that effective disaster-response planners and managers must look beyond geographical vulnerability in order to understand how unique social and political patterns in communities result in accentuated risk for some categories of people and how this can be averted.

Vulnerability is also closely linked to asset ownership. The level of resistance and resilience are largely determined by the assets and entitlements that individuals, households or communities can mobilize and manage during adversity. In general terms, the more assets people have, the less vulnerable they are, and also the greater the erosion of such assets, the greater the insecurity (Moser 1998, 3). Moser identifies important assets as labour, housing, social and economic infrastructure, household relations, and social capital. She argues that people do not simply draw upon their assets but possess sophisticated skills in utilizing them to cope with adversity and take advantage of opportunities (Moser 1998). Therefore, post-disaster recovery should not only aim at the provision of assets, as was mostly the case with the tsunami response, but it should also be backed with capacity and skills development to enable the affected people to utilize the assets to better their situation. Inevitably, interventions designed to build up the assets of the poor to withstand shocks will be increasingly important in reducing the human burden of hazards.

### *Humanitarian discourse and post-disaster recovery*

Ideally, humanitarian practice includes any activity aimed at amelioration of human suffering or improving the human condition, such as providing emergency assistance during a conflict, natural disaster or complex emergency. This has been and continues to be the major mandate of humanitarian agencies. However, the persistence of poverty amidst the ongoing humanitarian interventions justifies reviewing the traditional structural and systemic focus of the response to aspects of a potential or actual disaster. Indeed, some humanitarian agencies are gradually shifting the focus from just addressing the immediate effects of crises or disasters to dealing with the challenges which existed before the disaster event. They implicitly or explicitly embrace poverty reduction, recovery and long-term development.

Donor states also judge humanitarian action in terms of its political effects as much as its impact upon saving and protecting lives. The more inclusive nature of humanitarianism emanates from a growing belief that the value of saving lives only has a true meaning if those lives have a future that would spare them the deprivations that made them vulnerable in the first place (Macrae 2002). Thus, there is a growing interest in Disaster Mitigation and Preparedness (DMP), which is an indicator of a change in perception of disasters, with the old view of disasters as one-off events being replaced by an awareness of the connections between disasters and development processes.

Disasters are no longer seen as extreme events created entirely by natural forces but as manifestations of unresolved development challenges. This suggests that the extensive debate on the nature of vulnerability to disasters during the 1980s and early 1990s (Hewitt 1983; Blaikie et al. 1994) has found its way into humanitarian thinking. Unfortunately, these discussions have not fully been translated into comprehensive approaches to disaster management. For example, with the post-tsunami recovery, mitigation measures such as the buffer zone and warning systems have been established. However, the way in which they have been implemented (especially the buffer zone) has created more challenges than solutions, as will be discussed.

Often, disasters are reconsidered when they strike, attracting a sudden influx of funds. This is when pressure for DMP interventions emerges, but fades shortly after. Most of the funding is devoted towards meeting the short-term relief needs as opposed to long-term development. The rehabilitation and recovery phases attract insufficient funding and are usually considered the responsibility of the affected states (Walker et al. 2005). The short-term funding cycles and the complex interactions between vulnerability and broader development issues discourage concerted and consistent efforts (Christoplos 2003). With reference to the post-tsunami recovery, the generous funding has enabled the provision of relief as well as the development needs, as will be discussed in later sections of this paper. However, the major concern is whether these needs have been fully addressed.

### *Post-disaster recovery*

Post-disaster recovery presents a major challenge for humanitarian organizations, in that it is about moving towards development – a complex phenomenon described by Chambers (1997) as 'good change'. Haas et al. (1977) define disaster recovery as ordered, knowable and predictable. They identify four stages of activities: 1) emergency responses involving debris removal, provision of temporary housing, and search and rescue; 2) restoration of public services (electricity, water and telephone); 3) replacement or reconstruction of capital stocks to pre-disaster levels; and 4) initiation of betterment and reconstruction involving economic growth and development of the locale. All of these activities are part and parcel of the recovery process which can either occur simultaneously or in phases (Rubin 1985). However, recovery is context- and location-specific and defined by actions and aspirations, especially those of the affected people, during the shift from emergency and relief conditions. It involves a transition from rescuing survivors and providing relief to re-establishing basic services, infrastructure and economic activities, as well as the restoration of livelihoods. These are activities that have characterized the post-tsunami recovery in Sri Lanka, largely dominated by the reconstruction of infrastructure and restoration of livelihoods. Nevertheless, the process has not been fully guided by the aspirations of the affected people themselves, which has tended to compromise achievements.

The response to disasters is critical because in their wake disasters undermine previous development gains as well as compromise future gains, especially in poor countries. Blaikie et al. (1994) argue that susceptibility is often cyclical because disruption of assets and livelihoods by one event often make households yet more vulnerable to future hazards – the 'ratchet effect'. Hence, interventions must aim at mitigating the future impact of another disaster. Pelling (2003) has identified the new directions and theoretical approaches to disaster response, and has identified two schools of thought: the neo-Marxists and the behaviouralists. The neo-Marxist perspective considers disasters as deeply embedded within the social structures that shape everyday development experiences, and therefore any response must address such concerns (Hewitt 1983). In contrast, the behaviouralists argue that people only influence disasters in a limited way. They argue for policies focusing on disaster response and recovery. The entitlements perspective, widely adapted after the work of Blaikie et al. (1994), recognizes the interdependence of human and physical systems. Therefore, an effective response should consider both the human and physical aspects.

NGOs and humanitarian agencies play a major role in keeping acute human suffering on the global agenda both through early warnings about looming humanitarian crises and also through their presence and work in disaster-stricken areas. NGOs have been major advocates of better disaster preparedness, although whether they practice what they preach is yet to be explored. Many of the humanitarian agencies have signed onto the Sphere standards that seek to improve the quality of humanitarian response (Sphere 2004). Sphere is now firmly at the centre of humanitarian endeavours (Van Dyke & Waldman 2004). Its aims are to improve the effectiveness of humanitarian efforts and to enhance accountability within the humanitarian system. The Humanitarian Charter makes explicit links to the defined levels of service delivery set out in the five core sectors: water supply and sanitation; nutrition; food aid; shelter and site planning; and health services (Sphere 2004). These reflect the scope of humanitarian interventions in disaster-stricken places. Humanitarian agencies committed to the Charter and the Minimum Standards aim to achieve defined levels of service for people affected by calamity or armed conflict, and to promote the observance of fundamental humanitarian principles such impartiality and neutrality. This is critical in a politically and ethnically divided country such as Sri Lanka. However, priority is given to saving lives and addressing acute human suffering rather than to averting situations that create vulnerability. This is one of the major dilemmas of the aid industry.

It is important to confront questions of how to avoid or minimize the impacts of future disasters. However, addressing issues of sustainable vulnerability reduction requires political engagement an area loathed by humanitarians. Christoplos (2003) argues that risk management must be perceived as part of both development and humanitarianism. He argues that the risks that compromise development efforts must be addressed if development itself is to proceed as planned, and livelihood strategies should focus on addressing vulnerability rather than 'escaping' from poverty per se. Christoplos stresses that DMP should not be perceived as a project but rather as creating conditions for a broad-based 'culture of prevention'. This clearly captures the core debate of 'build back better'. In the next section, I will summarize the impact of the tsunami in Sri Lanka to set the context of the paper.

## The impact of the 2004 tsunami in Sri Lanka

After Aceh in Indonesia, Sri Lanka suffered the worst impact of the tsunami of 26 December 2004. The magnitude of the earthquake in the Indian Ocean, near the coast of Northern Sumatra, that triggered the tsunami was extremely high, measuring up to 9.0 on the Richter scale. It caused massive devastation. Out of a total of 25 districts, 14 were affected (more than 1000 km of the island's coastline). The most affected districts were those in the north-east. The overall impact is summarized in Table 1.

Approximately 150,000 people lost their livelihoods or employment, especially self-employed craftsmen (masons and carpenters) and fishermen. It is estimated that two-thirds of the island's fisheries sector was wiped out. Many fishermen and their families perished, while others were displaced. More than 65% of the fleet was destroyed or damaged including outboard motors, fishing gears and nets.

*Table 1.* The impact of *tsunami* in Sri Lanka

| Type of Damage | Total Numbers |
|---|---|
| Fatalities | 35,322 |
| Injured persons | 21,441 |
| Internally Displaced Persons | 516,150 |
| Affected districts | 14 out of 25 |
| Houses damaged or destroyed | 98,000 |
| Fishing boats damaged | 16,919 (75% of total fishing fleet) |
| Livelihoods affected | 150,000 |
| Health facilities | 97 |
| Educational facilities | 201, including 182 schools |
| Water and sanitation systems | 60,000 wells contaminated |
| Land affected | 9510 ha |

*Source:* Government of Sri Lanka (2005)

The damage to harbours and other related facilities was considerable. The tourist sector was also affected, resulting in further losses of jobs. The intrusion of seawater onto the mainland led to high levels of soil salinity, making fields unsuitable for cultivation and production. People also lost their domestic livestock and poultry.

In terms of infrastructure, the overall damage is estimated at USD 1 billion, which represents 4.5% of the country's gross domestic product (GDP). The Government of Sri Lanka estimated the cost of reconstruction and rehabilitation at approximately USD 2.2 billion, based on a more ambitious and comprehensive recovery and development programme that tackles problems and shortcomings that were existent in the pre-tsunami era (GoSL 2005a).

In the east and north, the affected population included people who had lived in camps as a result of the protracted civil conflict there. The ongoing war between the national government and its armed forces and the Liberation Tigers of Tamil Eelam (LTTE) has resulted in massive suffering for the affected communities. In these war-torn areas, the lack of resources, war-related destruction and 'brain drain' have resulted in shortages of skilled personnel in the key social sectors, thereby affecting the level of service delivery and the quality of life. The tsunami aggravated the situation of an already vulnerable population. In addition, restrictions and controls imposed by the army have created huge problems in obtaining regular and sufficient basic supplies and equipment, which in turn has also affected the recovery process. Humanitarian supplies are restricted in some of the uncleared areas (areas not under the control of the government and armed forces of Sri Lanka) which are still the site of active conflict and mines.

Many of the people in the conflict-ridden areas have already been victims of multiple displacements and as a result many have lost their livelihoods. Most of the communities in the conflict-affected areas ordinarily derive their livelihoods from agriculture or fishing, and the levels of poverty are higher compared to other parts of the country. The loss of income is attributed to a number of factors, which include: the death, disappearance and imprisonment of breadwinners; mass and protracted displacement; severe restrictions imposed by the military on work activities; presence of landmines and other unexploded ordinances on agricultural land; limited opportunities for work resulting from widespread destruction of industrial plants, the closure of ports and of other former sources of employment; and increased levels of poverty caused by the above factors which have resulted in the demise of local markets (Frerks & Klem 2005).

## Policy frameworks

In this section, I will examine the existing policy frameworks, both national and international, which were developed to guide post-tsunami recovery efforts. These form the basis for the discussion of the narrative 'build back better' (GoSL 2005a,b), which is the focus of this paper.

### *The guiding principles*

The Government of Sri Lanka (GoSL), international donors and civil society agreed to a set of core principles for tsunami recovery, which include:

(1) Resource allocation based on identified needs and local priorities, without discrimination on the basis of political, religious, ethnic, or gender considerations. There is need for sensitivity to neighbouring communities unaffected by tsunami.
(2) Subsidiarity – provides for locally appropriate solutions and allows a range of sub-national structures and organizations to be directly engaged in the recovery process. Recovery activities should be decentralized as much as possible, involving the lowest and most locally situated governance structures.
(3) Consultation with local affected communities and stakeholders. Interventions need to respond to clearly identified and articulated needs of the local communities on policy decisions.
(4) Communication and transparency in decision-making and implementation must be ensured at all levels. There is need for feedback to implementation authorities, grievance redress, transparent resource use and accounting, and zero tolerance of corruption.
(5) Future vulnerabilities ought to be reduced – the reconstruction processes should reduce future vulnerabilities to natural disasters through effective disaster management and early warning systems for possible future disasters.
(6) Analysis of individual interventions through conflict sensitivity (the distribution of aid across geographic regions and ethnic communities), sustainability, governance, gender, environment, land issues, and human rights concerns.
(7) Prudent management of debt relief.
(8) Coordination of the recovery effort in order to maximize benefits, prevent duplication, and minimize the burden on stretched government departments.

### *Presentation of the 'build back better' propositions*

In the aftermath of the tsunami disaster, Clinton (2006) provided ten propositions for defining 'build back better'. The slogan was widely adopted by different actors given the power of the author as the former President of the United States of America and the UN Secretary-General's Special Envoy for Tsunami Recovery. The propositions are as follows.

*Proposition 1: Governments, donors, and aid agencies must recognize that families and communities drive their own recovery.* This clearly articulates the agency of the affected individuals, families and communities. Clinton recommends full community participation through consultation in order to identify their needs and priorities and ensure local ownership. He also stresses the need for accountability to the affected people and recognition of the needs of different groups rather than having standard packages for all communities.

*Proposition 2: Recovery must promote fairness and equity.* This is very relevant in Sri Lanka's case, which is dominated by ethnic and religious divisions that pre-date the tsunami. It emphasizes addressing vulnerability and discrimination emanating from ethnicity and gender subordination. Clinton argues that 'build back better' would involve breaking inequitable development patterns in a sustained way.

*Proposition 3: Governments must enhance preparedness for future disasters.* Clinton stresses the need for robust systems that anticipate future disasters and measures that mitigate them. He recommends establishing government agencies and institutions, as well as setting laws and finances to address large-scale reconstruction needs with minimal delay and maximum effectiveness. He also argues for provision of life saving supplies and equipment.

*Proposition 4: Local governments must be empowered to manage recovery efforts, and donors must devote greater resources to strengthening government recovery institutions, especially at local level.* This stipulates the need to empower local governments to manage recovery efforts because they are uniquely placed to respond to the needs of individuals, households and communities. Clinton recommends their involvement in the planning, decision making, budgeting, and monitoring process at their level of jurisdiction. He highlights the need for investment in building local government and public service capacity.

*Proposition 5: Good recovery planning and effective coordination depend on good information.* Clinton emphasizes the importance of good information on damages, requirements and vulnerabilities in the recovery effort. He argues that proper needs assessment facilitates identification of existing opportunities for future development and charting a recovery strategy that goes beyond restoring what existed prior to the disaster. He recommends assessment beyond the resupply of equipment and reconstruction of infrastructure and institutions.

*Proposition 6: The UN, World Bank and other multilateral agencies must clarify their roles and relationships especially in addressing the early stage of a recovery process.* There is need for coordination with regard to division of labour between agencies and responsibilities with respect to financing and the nature of their partnerships.

*Proposition 7: The expanding role of NGOs and the Red Cross/Red Crescent carries greater responsibilities for quality in recovery efforts.* This highlights importance of the contributions of NGOs and the International Red Cross and Red Crescent Movement in the recovery efforts. Clinton recommends the following: building better partnerships between the various actors for sustainable recovery; the promotion of leadership among local communities and local aid groups; and strengthening local capacity to ensure accountability and coordination. He stresses the need for quality assurance measures and the promotion of optimal standards of professional conduct in their humanitarian response. This involves ensuring quality management, successful aid delivery, and engagement with local communities. He also highlights the need for informed and responsible giving from the donors, in which they must support actions that best reflect humanitarian and development practices in recognition that recovery is a long-term process. He urges NGOs to incorporate disaster awareness and the nature of effective and sustainable disaster preparedness.

*Proposition 8: From the start of recovery operations, governments and aid agencies must create the conditions for entrepreneurs to flourish.* This stresses the need for various actors to create opportunities for entrepreneurship in local communities. Clinton proposes the revival and expansion of private economic activity, employment, and securing diverse livelihood opportunities informed by economic realities for affected populations beyond the pre-disaster levels. Accordingly, recovery must encourage new investment as well as create more diverse and sustainable sources of income and not merely replacement of lost assets, social protection and temporary income generation. He also recommends the creation of an enabling environment for the development of markets and competitive advantage through appropriate policies and investment in infrastructure needed to stimulate trade and investment. In addition, governments and formal financial institutions need to make funding available as a way of reactivating small and medium enterprises, catalysing entrepreneurial energy and the creation of new micro-enterprises. He also recommends establishment of partnerships with national and international business actors to provide a source for technical inputs and financing in addition to enhancing access to markets for products.

*Proposition 9: Beneficiaries deserve the kind of agency partnerships that move beyond rivalry and unhealthy competition.* This calls for healthy competition and partnerships from the different actors in the recovery process. The synergies of the different capacities can be enhanced through partnership to facilitate resource mobilization, recovery strategies and division of labour.

*Proposition 10: Good recovery must leave communities safer by reducing risks and building resilience.* Ideally, any successful recovery effort must leave the survivors less vulnerable to future natural disasters. For example, schools and homes and critical infrastructure should be built to higher standards and on safer grounds, and there is also the need for early warning systems, education and public awareness on disaster reduction, and other risk mitigation measures. These must be backed by legal frameworks at national and local levels.

## *Unpacking 'build back better'*

Referring to Clinton (2006), Telford et al. (2006, 3) write:

> [O]ut of the rubble of the Indian Ocean's coastlines, and the suffering of its inhabitants, we would 'build back better': placing

coastal communities on a better development path; leaving survivors safer from future disasters; using the lessons learned today to ensure better responses in the future.

He emphasizes that the key test of a successful recovery effort is whether it leaves survivors less vulnerable to natural hazards. It is therefore vital that recovery plans identify the mechanisms that substantially reduce risk, and the resources for implementing such plans must be included from the outset. Successful risk reduction pays for itself many times over in the form of disasters avoided and lives safeguarded.

The guiding principles and the propositions share commonalities about the aims and purpose of the recovery process. For example, both emphasize the following: recognition of agency of the affected people; the need to promote equity and fairness; empowerment of the local authorities; community participation; transparency and accountability; partnership and coordination; and risk and vulnerability reduction. The propositions are more elaborate on how these can be achieved, while the guiding principles provide less detail on the modus operandi. However, it is also important to understand how other actors have interpreted and operationalized 'build back better'. In this respect, I will use information extracted from a workshop report. The theme of the workshop was '*Livelihood post-tsunami: Build back better?*' organized by the Institute of Policy Studies (IPS). The workshop[1] was attended by various stakeholders, including representatives from central and local government, civil society, donors, and the private sector, who were all working on post-tsunami livelihood issues (Institute of Policy Studies 2005a). In the following, I present some of the views on how they conceptualized the policy narrative in relation to livelihoods:

(1) Upgrading the key sectors such fisheries, tourism and small- and medium-scale entrepreneurs to achieve pro-poor growth
(2) Livelihood recovery perceived in the broader political, economic, social context, which entails not just overcoming the impact of the tsunami but also the legacy of the conflict and particularly the knock-on effect of the war upon the tsunami and efforts to 'build back better'
(3) Learning from and building upon the past experiences of poverty alleviation in Sri Lanka, particularly regarding what was successful and what kept people poor before the tsunami
(4) Using improved knowledge of demand and supply of livelihood activities, which involves understanding the complexity of livelihood strategies and the market situation. Interventions should shift from the supply side of production focus to demand-led and market-driven approaches.
(5) Strengthening the capacity of local governments, affected households, NGOs, and CBOs to plan, implement and monitor livelihood recovery programmes
(6) The need to provide a voice for affected households and communities
(7) Understanding of the local context regarding issues of ethnicity, castes, class, and gender, especially in the conflict-affected areas, and overcoming the inherent structural inequalities
(8) Overcoming mistrust and negative perceptions among key stakeholders. There is need both for improved information about interventions being undertaken by various stakeholders and better coordination.
(9) Improving the use of microfinance as a tool for poverty reduction.

According to the above conceptualizations, the 'build back better' narrative does offer the ideals for the recovery process, and it does cover most of the critical aspects that would improve the communities. However, whether these have been fully translated into practice is debatable. The narrative seems to promise more than it can actually deliver. A glance at the various interventions reveals the glaring gaps in the recovery process. The implementation has been an uphill task undermined by several factors, particularly the conflict that has left the country divided. In the south, the recovery process has been achieved to a great extent, whereas in the conflict-affected areas the recovery process leaves a lot to be desired. A lasting legacy of the post-tsunami recovery would entail reshaping of the long-entrenched divisions and bringing peace and reconciliation. All the conceptualizations emphasize drawing lessons from the past experiences and need for holistic interventions that address the pre-existing challenges. They argue for more people-centred approaches, the development of a multi-sectoral approach in planning and decision making, and improved access to resources. However, to gain a better understanding of the 'build back better' narrative it has to be considered in light of the actual interventions. The next section discusses the post-tsunami response and interventions in Sri Lanka.

## The response

The tsunami disaster attracted overwhelming assistance from corporations, governments, donors, and civil society (private individuals, clubs, and societies, local and international non-governmental organizations). The international community responded with humanitarian aid which included rescue teams, supplies, equipment, personnel, and funds. After the disaster, relief operations proceeded quickly and effectively, providing food, clean water, health care, and temporary shelters for hundreds of thousands of the affected people. Approximately USD 13.6 billion was raised for the tsunami relief and recovery operation, making it one of the best funded responses of all times. This is an equivalent of the annual global aid budget. According to Tsunami Evaluation Coalition (TEC), the total funding was equivalent to more than USD 7100 for every affected person (Telford et al. 2006). In Sri Lanka the total required relief, rehabilitation and reconstruction effort was estimated at USD 2.2 billion (GoSL 2005a). Today, the country is in transit from relief to recovery and development. Houses have been rebuilt, schools and health units reopened, livelihoods restarted, and people supported to deal with the psychological stress and trauma. The recovery process is now geared to long-term solutions such as housing, restoring livelihoods and diversifying the local economies. Donors and governments are also working to limit damage from future disasters and bolster their ability to respond. The United Nations and Indian Ocean countries have developed a regional tsunami early warning system. Environmental protection is also included in some recovery plans as a mitigation measure for future disasters.

In the next section I will examine selected interventions using some examples from the agencies I visited. However, many of the issues raised relate to the general humanitarian response and not specifically to these agencies. Nonetheless, I acknowledge many of the positive efforts and successes registered, especially those of individuals and organizations working under very challenging circumstances to alleviate the suffering of the affected populations.

## *NGOs in Sri Lanka*

Barely two days after the Indian Ocean tsunami struck Sri Lanka's coastline the international humanitarian agencies had started to arrive on the island en masse. A few, such as Oxfam, International Red Cross and Red Crescent Movement, Save the Children, and Care International came to augment pre-existing country programmes with additional emergency response capabilities. Since then, several humanitarian agencies have been working in Sri Lanka. Appendix 1 lists only the agencies working in Ampara District, but there are many other agencies operating throughout the country.

The number of international NGOs working in Sri Lanka increased from approximately 60 prior to the tsunami to more than 140 (Centre for Non Governmental Sector (CNGS) 2005), accompanied by an influx of resources and varying degrees of expertise. There was also a significant increase in the numbers of national NGOs and community-based organizations (CBOs), many of which support the INGOs in implementation of their programmes. Some of the newcomers had very limited knowledge about the context but came with 'precise and targeted' plans. They mainly engaged in the relief efforts without any comprehension of their impact on local dynamics, how they related to actual needs or whether they were duplicating the work of other agencies. Many of these vanished after the relief phase, while the more established organizations stayed on to address the reconstruction and recovery needs. The latter organizations received funding above the level of their annual budgets, which enabled them to venture into new programmes. Most of the funding that was received for the tsunami response has been channelled through non-governmental organizations and other humanitarian agencies. This partly explains why the focus of this paper is on the humanitarian actors. Given that every organization has its own mandate, it is vital to explore how their contributions relate to the narrative 'build back better'.

## *The interventions*

Many of the aid organizations clearly indicate in their mission statements the commitment to improving lives and reducing the vulnerability of marginalized groups. In this respect, the humanitarian agencies have played a critical role, ranging from the provision of emergency relief to recovery efforts focusing on the wider development needs in order to help people survive, rebuild and restore their livelihoods. The international NGOs were and still are engaged in various activities that include shelter, healthcare, education, gender advocacy, food security, and water and sanitation projects, as well as disaster preparedness and mitigation. Many of the organizations stress their commitment not only to building back what was damaged but also to responding to gaps in development and disaster preparedness so that the rebuilt communities are safer, more resilient and better able to respond to future natural hazards. Here, I will highlight a few selected interventions and examine how they relate to the 'build back better' strategy.

*Permanent housing.* Shelter has been one of the paramount humanitarian needs in post-tsunami Sri Lanka. An estimated 100,000 homes were destroyed or damaged by the tsunami. Getting people back into their homes is a major focus of the entire post-tsunami recovery. Government, donors, and international NGOs have committed funds to meet the housing needs of affected families. There has been significant progress on the housing programmes, especially in the south but the north and east are still lagging due to the conflict. According to the Reconstruction and Development Agency (RADA), as of the beginning of November 2006 more than 46,000 permanent houses had been built. Table 2 shows how in the south, Galle District and Matara District in the Southern Province, and also Kaluthara in the Western Province have fared very well, while in Hambantota, in the Southern Province, it is reported that the number of houses constructed is far in excess of the estimated requirement. Many of the new houses were built in the Sinhala areas in numbers greatly exceeding those actually destroyed by the tsunami. This clearly illustrates the ethnicity divide that predates the tsunami but which has undermined the recovery efforts in the minority groups. It is clear that resources allocation has not been based on the needs stipulated in the guiding principles and the related frameworks.

Many of the new houses have been provided with water and sanitation systems, drainage, access roads, and electricity. In the east there was a sharp contrast, with only a few new buildings that could be spotted as I travelled. In Ampara, many of the homes have basic structures of iron sheets, and brick and clay, compared to the state-of-the-art apartments elsewhere in the country. One of respondents expressed his dissatisfaction about the housing having heard about what the government was doing elsewhere. This particular respondent's house was mud-built and thatched with palm leaves. It was possible to see through the roof and he reported that whenever it rained the family gathered in one corner of the hut to avoid getting wet. Such houses are vulnerable to adverse weather and future natural events such as a tsunami.

With regard to housing, the major challenge has been the buffer zone restrictions which varied from region to region and been selectively applied. After the tsunami the Sri Lankan Government declared a no-build zone, 100–200 m

*Table 2.* Post-tsunami housing reconstruction

| Province & District | Number of houses required | Number of houses completed | Completed (%) |
|---|---|---|---|
| *Western and North Western Provinces* | *6600* | *919* | *14* |
| Colombo | 5639 | 347 | 6 |
| Gampaha | 887 | 498 | 56 |
| Puttalam | 74 | 74 | 100 |
| *Southern Province* | *33,917* | *34,972* | *103* |
| Hambanthota | 3193 | 5997 | 188 |
| Galle | 14,713 | 13788 | 94 |
| Matara | 8216 | 7791 | 95 |
| Kalutara | 7795 | 7396 | 95 |
| *Eastern Province* | *61,322* | *36,141* | 59 |
| Ampara | 28,349 | 16,067 | 57 |
| Batticaloa | 22,648 | 15,294 | 68 |
| Trincomalee | 10,325 | 4780 | 46 |
| *Northern Province* | *16,488* | *4555* | *28* |
| Jaffna | 9140 | 3380 | 37 |
| Mullaitivu | 5457 | 1032 | 19 |
| Killinochchi | 1891 | 143 | 8 |
| Totals | 118,327 | 76,587 | 65 |

*Source:* RADA (2007)

wide, along the coast, to limit casualties in any future tsunami. The Government also placed limits on agencies trying to rebuild according to the desires of the beneficiaries, since many people favoured revoking the buffer zone in order to return to their original land. The buffer zone policy received minimal stakeholders' support due its enormous social and economic impacts (Mulligan & Shaw 2007). It was a hastily devised policy that had significant implications for the already existent socio-economic disparities. Most of the humanitarian agencies had to abide by the policy when they embarked on the reconstruction. However, many agencies never fully grasped the implications of the buffer zone. People had to be relocated to marginal lands and far from their livelihood sources. The major challenge with the relocation of families from the buffer zone to other areas meant loss of livelihoods, particularly for those who depended on the sea. A survey by the International Labour Organization (ILO) found that 45% of fisher families who had been resettled 5 km from the coast had had been compelled to change their livelihoods (Mangrove Action Project 2007).

Although the aim of the buffer zone is to address vulnerability to the tsunami, it exposes people to more immediate risks. Negative impacts on equality, livelihoods and the environment (Oxfam 2005) resulting from the policy may increase the vulnerability of the affected populations in the long-term rather than decreasing it. Some of the new housing sites are considered not suitable by the inhabitants. A case in point is one of the housing projects visited in Hambantota, that was built on an elephant path and elephants have since damaged some of the houses and infrastructure, forcing people to abandon them. This has resulted in the wastage of resources that could have been used more efficiently if a proper assessment had been done. However, in contrast, some people expressed a fear of living close to the sea and would move if provided with *suitable* lands (Institute of Policy Studies 2005b).

In addition, the pre-existing challenges related to land, such as quality; availability and legal ownership, still prevail as well as the confusion of beneficiary eligibility and verification which have presented serious constraints not only to the humanitarian agencies but specifically to the beneficiaries. The ownership of the land is sometimes in dispute. The owners of houses which were located within the buffer zone have been resettled in the new housing projects (NHPs) located on state-owned land. These households are not required to demonstrate ownership of land and they also retain ownership of their former land in the buffer zone. However, in order to qualify for financial assistance, those with damaged houses outside the buffer zone have to document ownership (Center for Policy Alternatives 2005). Many of the tsunami survivors lost all of their possessions, including any documents that could establish their identities and hence many could not prove ownership that would make them eligible for assistance.

In some cases, women also experienced difficulties in accessing the housing grant that was mostly issued in the names of male household heads. Some reports indicate that the banks which administered payments were insisting that female property owners should sign their land over to their husbands in order to qualify for payment (ActionAid International 2006). In such cases, one wonders what happened to the widows and orphans. Another disadvantaged group was the tenants of the destroyed or damaged houses who did not qualify for housing assistance. This highlights the need for interventions to identify and target vulnerable groups, especially women and children who are supposedly 'represented' by men. It is important for development interventions to address the issues of power and powerlessness that make such groups vulnerable. Successful intervention must satisfy the questions of who participates in what and on what basis, the winners and the losers. A mere inclusion of the vulnerable is not sufficient; there is a need for this to be reflected in the development outcomes.

Kennedy et al. (2008) also raise key issues regarding 'build back better' in relation to settlement and shelter. They argue that success was mainly gauged by the number of houses constructed as opposed to the quality. It is argued that many of the organizations had limited knowledge and experience in construction, coupled with a race against time to meet the funding deadlines and beneficiary expectations. This meant that the focus on safety, security and livelihoods was lost. Furthermore, there was a desire for 'modern' dwellings which represent affluence for the middle-class as opposed to traditional dwellings that possess seismic-resistance properties. Western-style concrete structures were requested as being representative of 'development' and progress' even though they are less suited to the climate and do not utilize the relevant traditional knowledge about settlements and shelter.

Many of the housing units have also been designated as culturally inappropriate in terms of ensuring privacy, and the housing design was alien to the normal practices. For example, the ventilation systems in some of the housing schemes were designed on the assumption that gas or electricity would be used for cooking, despite the widespread use of cheaper biomass fuels such as wood and paddy husks (Institute of Policy Studies 2005b). While it can be argued that this is a move for the better given the impact of biomass fuel on the environment, the cost of gas and electricity is prohibitive for many rural families.

*Health and care.* the tsunami caused significant damage to health infrastructure resulting in reduced quality of health services and access to health facilities. Some humanitarian agencies are working with the Ministry of Health to reconstruct, rehabilitate or upgrade the health units in Sri Lanka. In addition, they are procuring urgently needed medical equipment, fostering well-being through health awareness and facilitating capacity development through the construction of medical training schools. However, the need for improved health services goes beyond the tsunami-affected coastline. A number of areas have limited access to health personnel and facilities especially in the war zones. Under such circumstances 'build back better' would imply the provision of infrastructure, equipment and personnel, especially in the conflict-affected areas where there are glaring gaps in service delivery worsened by the disappearance of skilled personnel. Some of the beneficiaries visited reported that they still had to travel more than 18–25 km in order to access health care. In addition, although free healthcare is provided at the government health units, transport costs are very prohibitive. For example, one of the respondents reported spending 2000 LKR (USD 18) for every trip, which was equivalent to his monthly income. He had to make frequent trips to the hospital due to the nature of his illness that required regular medical attention. Despite the significant improvement in social services, particularly in terms of infrastructure and equipment and supplies, the issue of personnel is still a challenge. Without personnel, the challenges of poor service provision remain and this may compromise the quality of health care received. There is need to work closely with government, which has the responsibility of training and deploying personnel.

*Water and sanitation.* Many organizations have been involved in ensuring access to clean water through the production and distribution of drinking water as well as the construction and upgrading of wells and latrines. For example, the International Federation of Red Cross and Red Crescent Societies set up water treatment plants, storage tanks and a mechanism for water tanking in affected areas. In addition, they introduced solar-operated wells, and rainwater harvesting in communities and schools, including local production and distribution of ceramic water filters to ensure supplies of safe drinking water. This has greatly contributed to meeting the needs for clean water, which in turn minimizes the risk of water-borne diseases. However, in Ampara District some of the households I visited reported a lack of latrines and access to water for domestic use. It was reported that family members had to travel to irrigation channels in order to take a bath and even farther to access clean water for domestic use. The biggest challenge lies in longer term investment in the development of water and sanitation infrastructure, particularly in the dry zones and coastal areas, as well as the wider sensitization to sanitation and hygiene practices.

*Livelihoods.* Approximately 150,000 people lost their livelihoods as a result of the tsunami (GoSL 2005a). Coastal communities faced severe economic stress in the aftermath of the tsunami. Most of the affected families depended on informal sectors for their livelihoods, such as fishing, tourism, cottage industries, petty trade, and wage labour. Many families were affected through the loss of breadwinners, collapse of markets, and the destruction of assets and infrastructure. The restoration of livelihoods has been a major component of post-tsunami reconstruction and recovery. According to the International Federation of Red Cross and Red Crescent Societies, the support for livelihoods is aimed beyond the replacement of lost income-generating material assets and encompasses skills development and vocational training for the affected populations. In many cases, the restoration of livelihoods has been tied to the housing programme, whereby the beneficiaries of housing have also been registered for the livelihood scheme. This is a more holistic approach to recovery that provides a 'starter kit' catering for both the livelihoods and housing needs for the affected families.

Many fisher families and cooperatives received motorized boats, engines and nets, while women have benefitted from targeted distributions of sewing machines, coir making machines and other home-based livelihood assets. Some programmes targeted specifically women, for example training and assistance in rapid income-generation projects such as food processing, agriculture, livestock, and home-gardening (IFRC 2006). Other activities targeted both men and women, and included mushroom and spice cultivation, horticulture and small-scale home-based food processing projects. Skills development has focused on providing training in tile laying, masonry and carpentry for construction workers to meet the surge in the booming construction industry. Young people are also participating in computer training and English language classes to improve their job prospects (IFRC 2006).

From the interventions, it is clear that an attempt is being made to build better livelihoods through skills development, improved technology and focus on the most vulnerable groups. However, most of the interventions have focused heavily on asset replacement with a bias towards the fisheries sector. According to RADA, of the 19,000 fishing boats destroyed, 90% have been replaced (RADA 2006). Many organizations were involved in boat distribution, leading to an oversupply of boats. Several cases are cited where there was indiscriminate distribution even the non-fisher folk received fishing equipment yet others who had genuinely lost their boats did not receive any replacement (Creech 2005). In addition, some households managed to obtain as many boats as they could while others completely missed out.

This intervention may also have assumed homogeneity of fishing villages, that all the coastal communities had a shared interest in fishing. Given that many of the boats are now parked next to people's homes there could have been a better way of identifying families whose livelihoods are based on fishing and the dynamics of fishing crews. For example, 'community'-based rules also influence access to resources and therefore not everyone with fishing gear can venture into fishing. Social relations such as ethnicity or caste determine who can access fishing opportunities. In addition, many families engage in diverse activities in order to achieve livelihood security and they devote their energies to the more familiar and most productive activities. Hence, in addressing vulnerabilities there is a need to learn from communities and households themselves what activities they value most in rebuilding their livelihoods, as a way of maximizing scarce resources. Many of the livelihood interventions were standardized and not addressing the specific needs of regions and communities. It was only at a later stage that some organizations ventured into conducting proper needs assessments. However, it was difficult to establish whether the interventions met the needs since the exercises were still ongoing and implementation was pending.

Furthermore, the provision of more efficient fishing gear compared to the traditional gear presents its own future challenges. Improving the efficiency of fishing efforts definitely serves the purpose of securing increased incomes for fishing families. However, it may present a long-term danger of over-exploitation and depletion of the resource on which many depend on for their livelihoods. According to the humanitarian news and analysis agency IRIN (Integrated Regional Information Networks), fishermen are catching 70% of what they did before the tsunami. The agency reports that in southern Sri Lanka fishermen are now experiencing the effects of overfishing (IRIN 2007). This may become a future avenue for vulnerability unless it is combined with policies aimed at regulating catches. In addition, the maintenance of modern equipment has cost implications for fisherfolk in terms of fuel, maintenance and repairs. This may explain why some of the boats have not served their intended purpose because the maintenance costs exceed the benefits, and hence the opportunity for enhanced livelihoods is lost.

Oxfam GB is supporting coir workers to improve access to markets and add value to products. The coir workers were affected by the tsunami when it destroyed their production sites and equipment. The intervention was designed to rehabilitate and restore the coir livelihood sector through the provision of cash grants. The cash grants were provided to facilitate cleaning of the husk pits, and purchase of coconut husks and yarn spinning wheels. However, Oxfam GB realized that restoration of lost equipment was not sufficient. For this reason, it has supported quality improvement of the products as well as facilitating access to market through links to the potential buyers. Sustainable production requires a matched market demand for the products. In the case of coir production, significant progress has been made in boosting production beyond the pre-tsunami levels. However, it was reported that the quality of the products is still poor and thus not competitive on the market. In addition, it was reported that the local industry faces competition from the influx of imported goods of superior quality. Domestic rivalry from similar projects has also resulted in competition for markets, thus driving the market prices down and compromising the goal of increased income.

The provision of cash grants is also one of the major interventions in the livelihoods restoration. In Ampara District, I visited some of the families that had or were to receive cash grants from the International Red Cross and Red Crescent Movement aimed at improving their livelihoods. They were to receive cash grants amounting to 15,000 LKR (USD 139) per household, given in two instalments. Some families managed to utilize the grant to boost their agricultural production. However, for others the intervention has not worked as intended. One of the respondents, an old man in his early seventies, with a frail and ill wife, had also received the first instalment of the grant. He used the grant to meet the medical expenses of his wife and had not embarked on the planned activities. Treating his wife was a priority, despite the poverty. Such cash transfers, without stipulation on the ways in which the money should be spent, mark a new and fashionable trend in aid circles. In such cases, the pre-existing conditions of the beneficiaries are critical when planning any interventions. It appears rational to first attend to the immediate and urgent needs before embarking on the long-term ones. Special consideration should be made for such vulnerable groups to enable them to benefit from the interventions.

Overall, the livelihood interventions are crucial for the recovery process and significant progress has been registered. Available information suggests that 70–85% of the households affected by the tsunami had regained their main source of income as of November 2005 (GoSL 2005a). However, to ensure better outcomes there is need for planned asset replacement that considers market conditions, and the capacity and experience of the recipients. In addition, the promotion of alternative livelihoods was largely ignored. Many of the interventions focused on restoration of traditional occupations. This has it own advantages, especially in increasing output. However, diversifying the old forms of livelihood pursuits is necessary, especially those that are no longer appropriate and productive. The new livelihood opportunities must match the market trends.

*Agriculture.* Supporting agricultural initiatives in order to ensure food security and livelihoods has also been one of the major interventions. For example, Oxfam GB in Hambantota is supporting small- and medium-scale farmers through improved production techniques, diversification of crops, negotiations for land entitlements, and market chain improvements. Oxfam GB supported the construction of a cattle and elephant fence in the small village of Julpallama in Tissamaharama Division (Southern Province) as a way to protect the agricultural land from elephant invasions which had been a long-standing problem in the community. In addition, it supported agricultural development through the restoration of feeder roads and provision of specific crops valued by farmers, such as green gram (mung bean), groundnuts, and *kiriala yam* (*Colocasia* species). It has also supported one of the critical demands for agricultural development in the Dry Zone through the establishment of irrigation systems and increased capacity of water reservoirs. In other areas, Oxfam GB is supporting the installation of irrigation systems as well as linking farmers to providers of extension services and sourcing markets for their produce. This is also a comprehensive approach to enhancing agricultural production which is a major livelihood for people in this area. The organization is directly responding to the community's priorities, according to some beneficiaries. The cattle and elephant fence was a purely a community initiative and the organization only provided the financial support. This is clearly one of the ways that the organization has been able to tap people's agency and hence meet actual needs. The people have demonstrated the ability to mobilize for their own development. The community also set up a fund for the maintenance of the solar-powered fence.

However, the issue of land was highlighted by the beneficiaries as a major challenge. Land is considered as the single most important asset and the lack of freehold rights is attributed as a major cause of poverty. In Tissamaharama, the area that I visited, my respondents reported the lack of full rights to the government land. Some had title deeds that enabled them to pass on the land as inheritance to their children. However, these accounted for only a few, while many had permits that only granted partial rights such as settlement and cultivation but not inheritance. According to the respondents this deters more permanent developments despite the minimal risk of eviction. Furthermore, those with title deeds reported that the land was not valued and therefore the deeds could not be used to obtain mortgages or to secure loans. Their major concern was that they would like to use the land as collateral for securing loans in order to boost their agricultural activities and other businesses as well as pass on the land to their children or sell it if they wished to relocate. Legal issues concerning land tenure were therefore shown to be a central and unresolved issue. These issues require policy which involves engaging policy makers in re-examining the land tenure and ownership system. Oxfam GB reported that it was engaged with the concerned authorities in trying to resolve the problem. It was also reported that there are many squatters (cultivating without permission), which could result in the exploitation of such farmers by the bona fide occupants in terms of labour and other remittances that would compromise their livelihood outcomes. The longer term solution would be to review the tenure systems and regularize ownership.

*Disaster preparedness and management.* A few organizations are engaged in supporting efforts to build local capacity through the introduction of a community-based approach aimed at strengthening the coping mechanisms of communities by identifying threats and developing collective response plans to warn against and respond to future hazards. Other initiatives include support for a schools-based programme which involves training teachers to educate children about hazards and how to respond to them. It entails training and provision of essential equipment and tools to be used in the management of disasters. In contrast, other programmes have focused on training their staff to better identify post-disaster needs and to manage effective and efficient disaster response operations.

However, a well-integrated disaster management system is still weakly implemented. This should be a cross-cutting theme in all interventions and deliberate effort should be made to increase the resilience of affected communities in the case of future disasters, for example vulnerability mapping crucial to improve needs assessment, the establishment of savings at local level and national reserves, empowering local communities, housing safety, and environmental conservation. This may serve to minimize destruction and avail resources required for future disaster responses which are increasingly diminishing amidst competing demands.

## Constraints to achieving 'build back better'

Overall, from the above discussion there are emerging issues that have compromised the post-tsunami recovery effort and these will be briefly discussed here.

The sudden increase of International NGOs with approximately 100 newly registered after the tsunami (Centre for Non Governmental Sector (CNGS) 2005) presented challenges in terms of competition for 'clients' and uncoordinated interventions leading to duplication of efforts and wastage of resources. For example, the oversupply of assets highlighted above. In addition, the efficiency and quality of NGO programmes was compromised by lack of experience and pressure for quick results (Haug & Weerackody 2005). The aid system does not fully comprehend the challenges of reconstruction and recovery. Furthermore, many of the organizations are funded to provide specific services irrespective of community needs. The disparity between community needs and the mandates of organizations compromises well-intentioned interventions.

In addition, the earmarking of funds for specific purposes and places was detrimental to addressing issues of equity. Many of the donors provided funds specifically for the tsunami-affected communities. There was little consideration for the conflict-affected areas or other poor areas equally in need of assistance. However, some humanitarian agencies, such as Oxfam, have been flexible in the nature of their funding. In the Oxfam South Programme deliberate effort has been made to support the indirectly affected but vulnerable communities such as those visited for this research.

When travelling from the southern to the eastern part of the island it was clear that there are marked contrasts in the levels of development. The actual pattern of tsunami response aggravated these disparities as more resources were channelled to the south. In the east and north, the conflict has profoundly reduced opportunities for development offered by the tsunami recovery effort. The situation is characterized by fear and uncertainty which affects long-term planning. Many humanitarian agencies have suspended their operations in the north and east and this implies that these communities have lost in terms of the tsunami funding which is time-bound except for some agencies such as the IFRC, which has a frozen fund for the conflict-affected areas. Of course, 'build back better' would mean resolving the conflict which has instead worsened since the breakdown of the ceasefire in 2006. Although this is not directly attributed to the work of NGOs, to some extent there has been increased tension stemming from aid resources allocations that have continued to favour the Southern Province over the other provinces. Earmarking aid is an issue that has to be revisited in the aid system to enable humanitarian organizations to reach out to all deserving populations.

The aid system also creates pressures to design work within project structures that single out limited sets of activities within a specific time frame. Many of the organizations were under the pressure to spend within the limited time frames. This also immensely affected the quality of the different interventions and outcomes, as discussed earlier. There are reports of funding shortages for the much needed housing and other programmes. Short-term aid is not sufficient for creating self-reliance and there is a need in the longer term for creating safety nets for dealing with the chronically poor and for those shocked by disaster (Sumarto et al. 2000; Devereux 2002).

Both the guiding principles and the propositions emphasize the need for consultation and participation by the communities in the reconstruction planning and implementation. However, the extent to which this was pursued by the various actors is limited (Uyangoda 2005; Weinstein 2005). Participation was determined mainly by organization principles and policy as well as government procedures for beneficiary selection. Some organizations had preconceived ideas of what they wanted to achieve in the affected communities. This was a major weakness and compromised the achievements. In addition, the pressure to demonstrate quick results did not allow for thorough consultations with the affected communities. The outcomes of the interventions clearly illustrate the importance of thorough consultation. For example, earlier, people had never been consulted regarding the quality, size and location of their new homes which to a large extent affected their livelihoods. In many cases, there were standardized packages that did not cater for individual needs (the example of non-fisher folk receiving fleets of boats is a case in point).

Political factors have also clearly played a critical role in the recovery process. Political interference is a key constraint in equitable allocation of the post-tsunami resources. For example it is documented in Hambantota, the constituency of President Mahinda Rajapakse, that the housing programme boasts a completion rate of 173% compared to Ampara, where the completion rate is 39% (RADA 2006). Most of the construction work and interventions in general to date have been concentrated in the southern districts while there has been little progress in the severely affected Ampara, Batticaloa and Trincomalee Districts in the Eastern Province. Korf (2004) also emphasizes the role of social and political assets in enabling individuals, households, or economic actors to stabilize or even expand their livelihood options and opportunities. The role of political patronage in Sri Lanka ultimately shapes access to resources. The over-provision of housing in the south clearly indicates an inequitable allocation of resources from a national perspective. The ethnic segregation has not been based on individuals per se but is manifest in the regional imbalance in resource allocation.

Poor targeting is one of the major constraints in the recovery efforts. The eligibility of many of the recipients has been questioned (Haug & Weerackody 2005). Some of the beneficiaries were not genuinely affected while the most deserving were excluded. There was need for more transparency in the selection of beneficiaries to provide opportunities for the most disadvantaged rather than rewarding the well-to-do who benefitted from better connections. The Divisional Secretariats (administrative units) played a crucial role in the beneficiary identification and selection. In applying the principle of subsidiarity or decentralization of aid this was the way to go. However, the beneficiary selection has been highly criticized because the process of compiling beneficiary lists was very political. The partisan compilation was aimed at rewarding their supporters who were not necessarily the victims (Mulligan & Shaw 2007).

Despite the ongoing efforts, the deficiencies in targeting the provision of support have combined with other factors to exclude the most vulnerable groups. The attempts by humanitarian organizations that have tried to involve women in many of the interventions are commendable. With regard to ethnicity, beneficiaries from the different ethnic backgrounds are involved. In Ampara, I visited one of the women's community-based organization (CBOs) where women from the different ethnicities and religions were working together harmoniously. The management committee of the CBO had representatives from different ethnic groups. However, during the interviews questions relating to ethnicity were avoided for ethical reasons. At the national level, the discussions with the agencies' senior officials highlighted the commitment to work with all groups, irrespective of gender or ethnicity. Many organizations have made a deliberate effort to allocate resources to all the affected regions where conditions permit. Nevertheless, the south has received the lion's share in the recovery efforts while the most deserving groups have been excluded or even left in a worse position than prior to the tsunami.

Poverty, in terms of both numbers of people and severity, has increased after the tsunami, as revealed in a study titled *People's Verdict on Tsunami Recovery in Sri Lanka* (Sarvananthan 2007). The household income data from the survey indicates that the majority of the tsunami-affected households live in absolute poverty. According to Sarvananthan (2007), the poverty levels in the south, north and east had risen from 64% to 80% after December 2004. The survey reveals that 57% of the tsunami-hit families earned less than

USD 1 per day, the internationally accepted poverty line (Inter Press Service 2007). More than 30,000 affected people registered with RADA their complaints, wishes and aid requests regarding their livelihood recovery. This is a clear indication that two years after the tsunami devastation a sizeable number of the affected people are still languishing without support to rebuild their livelihoods despite massive interventions.

## Conclusions

Hilhorst (2004) argues that disasters such as the 2004 tsunami in Sri Lanka can either enhance radical change or bureaucratic reform or reinforce existing power relations. The analysis of vulnerability and events that occurred after the tsunami demonstrates to a large extent a return to the status quo in Sri Lanka. For some, it has been a process of radical change but for many, especially in the northern and eastern parts of the island, little has changed.

Post-disaster reconstruction goes beyond the construction of new homes because '[without] an understanding of the local, political and cultural complexities' attempts to resolve any problems will be difficult (Macrae & Zwi 1992, 300–301). 'Build back better' would mean dealing with the risk factors that compromise livelihoods as well as ensuring increased coping capability of individuals and communities. As highlighted above, the interventions by humanitarian agencies try to address these key aspects, although not totally because many of the external risk factors are beyond their scope and need collaborated efforts with other actors, particularly the Government of Sri Lanka.

Vulnerability reduction is critical as a way of averting or minimizing the impact of future disasters. There is need to identify and address the sources of vulnerability for better recovery outcomes. Today, when aid is increasingly given for disaster response as opposed to 'ordinary development', disasters present opportunities to address the pre-existent development challenges of affected communities. For Sri Lanka, the generous funding created such an opportunity that has not been fully harnessed. Post-disaster recovery should not be simply seen as replacing 'lost' assets but empowering communities and governments to redefine the course of the country's development. This entails giving people the capability to engage fruitfully and meaningfully with the new circumstances and, most importantly, the capability to change their circumstances. These assets should ultimately enable affected communities to act and reproduce, and challenge or change the rules and structures that govern the control, use and transformation of resources. Given the current status of the recovery effort, 'build back better' has not lived up to expectations. The structural vulnerabilities are still evident and even deeper than before.

In the past, post-crisis situations have been a turning point in the course of many countries' development. For example, many of the European countries that were supported under the Marshall Plan after World War II managed to rebuild their economies far beyond pre-war levels. There is need to borrow lessons of what worked in the past and why it worked, and these can be contextualized for the different situations.

## Note

1 Held 1–2 December 2005, at BMICH, Colombo.

*Acknowledgements.* – I am very grateful Professor Ragnhild Lund for her insightful inputs and assistance rendered during the fieldwork in Sri Lanka, and the GAP Project members, especially Professor Piers Blaikie who provided useful feedback for the article. I also extend my gratitude to the staff at the FONT office for their facilitation. Special thanks are due to the officials at the International Committee of the Red Cross and Oxfam GB for permitting me to study their organizations and for all the assistance rendered.

## References

ActionAid International. 2006. *Tsunami Response: A Human Rights Assessment*. ActionAid International, New Delhi.

Anderson, M.A. 1994. *Understanding the Disaster-Development Continuum: Gender Analysis is the Essential Tool*. Walker, B. (ed.) *Women and Emergencies*, 7–10. Oxfam, Oxford.

Blaikie, P., Cannon, T., Davis, I. & Wisner, B. 1994. *At Risk: Natural Hazards, People's Vulnerability, and Disasters*. Routledge, London.

Bolin, R.C. & Stanford, L. 1998. The Northridge earthquake: Community-based approaches to unmet recovery needs. *Disasters* 22:1, 21–38.

Centre for Non Governmental Sector (CNGS). 2005. *NGOs*. http://cngs.erd.gov.lk/ngo.html (accessed September 2007).

Centre for Policy Alternatives (CPA). 2005. *Landlessness and Land Rights in Post-Tsunami Sri Lanka*. http://www.cpalanka.org/research_papers/IFRC_land_study.pdf (accessed August 2008).

Chambers, R. 1997. *Whose Reality Counts?: Putting the First Last*. Intermediate Technology Publications (ITP), London.

Christoplos, I. 2003. Actors in risk. Pelling, M. (ed.) *Natural Disasters and Development in a Globalizing World*, 95–109. Routledge, London.

Clinton, W.J. 2006. *Lessons Learned from Tsunami Recovery: Key Propositions for Building Back Better*. United Nations Secretary-General's Special Envoy for Tsunami Recovery. United Nations, New York.

Creech, S. 2005. Tsunami issues affecting fishing communities and the challenges to be addressed if 'build back better' is to contribute toward sustainable livelihood development. Steele, P. (ed.) *Livelihoods in Post-tsunami Sri Lanka: Building Back Better?*, 43–63. Institute of Policy Studies, Colombo.

Devereux, S. 2002. *Social Protection and the Poor: Lessons from Recent International Experience*. IDS Working Paper No. 142. University of Sussex, Brighton.

Flint, M. & Goyder, H. 2006. *Funding the Tsunami Response*. Tsunami Evaluation Coalition, London.

Frerks, G. & Klem, B. 2005. *Tsunami Response in Sri Lanka, Report on a Field Visit from 6–20 February 2005*. http://www.tsunami-evaluation.org/NR/rdonlyres/7721073D-0868-410C-AF26-200165D387F2/0/Wageningen_tsunami_response_in_srilanka_20050314.pdf (accessed July 2007).

GoSL. 2005a. *Sri Lanka: Post Tsunami Recovery and Reconstruction. Joint Report of the Government of Sri Lanka and Development Partners*. Government of Sri Lanka, Colombo.

GoSL. 2005b. *Government of Sri Lanka Post Tsunami Reconstruction and Recovery Strategy*. Government of Sri Lanka, Colombo.

Haas, J.E., Kates, R.W. & Bowden, M.J. (eds.) 1977. *Reconstruction following Disaster*. MIT, Cambridge, MA.

Haug, M. & Weerackody, C. 2005. Delivery of tsunami aid for livelihood development at the community level. Steele, P. (ed.) *Livelihoods in Post-tsunami Sri Lanka: Building Back Better?*, 19–33. Institute of Policy Studies, Colombo.

Hewitt, K. 1983. *Interpretations of Calamity: From the Viewpoint of Human Ecology*. Allen Unwin, London.

Hilhorst, D. 2004. Complexing and diversity: Unlocking social domains of disaster response. Bankhoff, G., Frerks, G. & Hilhorst, D. (eds.) *Mapping Vulnerability, Disasters, Development and People*, 52–66. Earthscan, London.

IFRC (International Federation of Red Cross and Red Crescent Societies). 2006. *Two Years on: Post Tsunami Recovery Sri Lanka 2006*. IFRC, Colombo.

Institute for Policy Studies (IPS). 2005a. *Livelihoods Post-tsunami: Build Back Better?* http://www.ips.lk/news/newsarchive/2005/01122005_p_tsun/workshop_report.pdf (accessed May 2008).

Institute for Policy Studies (IPS). 2005b. *Listening to Those Who Lost: Survey and Analysis of Rebuilding and Relocation of Tsunami Households in Sri Lanka*. Institute of Policy Studies, Colombo.

Inter Press Service. 2007. *Sri Lanka: Tsunami Recovery Skewed by Sectarian Strife*. Amantha Perera. 3 January 2007. http://ipsnews.net/news.asp?idnews =36051 (accessed June 2008).

IRIN. 2007. *Sri Lanka: Fishing Industry Still Lacks Large Tonnage Boats*. 21 June 2007 http://www.irinnews.org/report.aspx?ReportId=72857 (accessed June 2008).

Kennedy, J., Ashmore, J., Babister, E. & Kelman, I. 2008. The meaning of 'build back better': Evidence from post-tsunami Aceh and Sri Lanka. *Journal of Contingencies and Crisis Management Humanitarian Exchange* 16:1, 24–36. http://www.blackwell-synergy.com/doi/abs/10.1111/j.1468-5973.2008.00534.x (accessed May 2008).

Korf, B. 2004. War, livelihoods and vulnerability in Sri Lanka. *Development and Change* 35, 275–295.

Lewis, J. 1999. *Development in Disaster-prone Places*. IT Publications, London.

Macrae, J. (ed.) 2002. *The New Humanitarianism: A Review of Trends in Global Humanitarian Action*. Humanitarian Policy Group Report 11. Overseas Development Institute, London.

Macrae, J. & Zwi, A. 1992. Food as an instrument of war in contemporary famines: A review of the evidence. *Disasters* 16:4, 299–321.

Mangrove Action Project (MAP). 2007. *Sri Lanka Buffer Zone Forces Fisherfolk into Other Work: Survey*. Niranji Jayawardena, 12 September 2007. http://www.mangroveactionproject.org/news/current_headlines/sri-lanka-buffer-zone-forces-fisher-folk-into-other-work-survey/?searchterm=lankan (accessed June 2008).

Morrow, B.H. 1999. Identifying and mapping community vulnerability. *Disasters* 23:1, 1–18.

Moser, C. 1998. The asset vulnerability framework: Reassessing urban poverty reduction strategies. *World Development* 26:1, 1–19.

Mulligan, M. & Shaw, J. 2007. What the world can learn from Sri Lanka's post-tsunami experiences. *International Journal of Asia-Pacific Studies (IJAPS)* 3:2, 65–91.

Oxfam. 2005. *Targeting Poor People. Rebuilding Lives after the Tsunami*. Oxfam, London.

Pelling, M. 2003. *Natural Disasters and Development in a Globalizing World*. Routledge, London.

Quarantelli, E.L. 1987. What should we study? Questions and suggestions for researchers about the concept of disasters. *International Journal of Mass Emergencies and Disasters* 5, 7–32.

Quarantelli, E.L. 1995. What is a disaster? *International Journal of Mass Emergencies and Disasters* 13:3, 221–230.

RADA. (Reconstruction and Development Agency). 2006. *Post Tsunami Recovery and Reconstruction*. Mid Year Review, June 2006. Reconstruction and Development Agency, Colombo.

RADA. 2007. *Livelihood Newsletter*. January – February 2007. http://www.humanitariansrilanka.org/Livelihood/PDFs/Resources/RADA%20Newsletter-January-February%20 2007.pdf (accessed October 2008).

Rubin, C.B. 1985. The community recovery process in the United States after a major natural disaster. *International Journal of Mass Emergencies and Disasters* 3, 29–28.

Sarvananthan, M. 2007. *People's Verdict on Tsunami Recovery in Sri Lanka*. International Centre for Ethnic Studies, Colombo.

Sphere. 2004. *Sphere Humanitarian Charter and Minimum Standards in Disaster Response*. Sphere Project, Geneva.

Sumarto, S., Suryahadi, A. & Pritchett, L. 2000. *Safety Nets and Safety Ropes: Who Benefitted from the Two Indonesian Crisis Programmes – The 'Poor' or the 'Shocked'?* Policy Research Working Paper No. 2436. World Bank, Washington, DC.

Telford, J., Cosgrave, J. & Houghton, R. 2006. *Joint Evaluation of the International Response to the Indian Ocean Tsunami: Synthesis Report*. Tsunami Evaluation Coalition (TEC), London.

Uyangoda, J. 2005a. Post-tsunami recovery in Sri Lanka. *Polity* 2:3, 4–7.

Van Dyke, M. & Waldman, R. 2004. *The Sphere Project Evaluation Report*. http://www.sphereproject.org (accessed August 2008).

Varley, A. (ed.) 1994. *Disasters, Development and Environment*. John Wiley, Chichester.

Walker, P., Leaning, J. & Minear, L. 2005. Smoke and mirrors: Deficiencies in disaster funding. *BMJ* (British Medical Journal) 330, 29 January 2005, 247–250.

Weinstein, H.M. 2005. Sri Lanka. Fletcher, L.E., Stover, E. & Weinstein, H.M. (eds.) *After the Tsunami: Human Rights of Vulnerable Populations*, 57–73. University of California, Berkeley.

Wisner, B., Blaikie, P., Cannon, T. & Davis, I. 2004. *At Risk: Natural Hazards, People's Vulnerability and Disasters*. Routledge, London.

# Appendix 1

Humanitarian agencies operating in Ampara District

| | |
|---|---|
| AAI | ActionAid International |
| ACF | Action Contre La Faim |
| ACYMMAC | All Ceylon Young Men's Muslim Association Conference |
| ADT | Alliance Development Trust |
| AFMAL-FBF | Associazione Con I Fatebenefratelli Per I Malati Lontani |
| Ai.Bi. | Amic' Die Bambini |
| AoG | Apotheker Ohne Grenzen Deutschland e.V |
| ARCHE | arche noVa e.V. Dresden |
| Architecture for Humanity | |
| ASIA RELIEF | Asia Relief |
| BAJ | Bridge Asia Japan |
| BCS | Brothers of Charity Sri Lanka |
| BF | Berendina Foundation |
| BRC IFRC | British Red Cross |
| CA | Christian Aid UK |
| CA | CORDAID |
| CAM | Comite d'aide Medicale |
| CARE | CARE International |
| CARITAS | Caritas |
| CBM | Christian Blind Mission |
| CCF | Christian Children's Fund |
| CHA | Consortium of Humanitarian Agencies |
| Charity Trust for Advanced VCT Centres | |
| CHF | CHF International |
| CIDA | Canadian International Development Agency |
| CMSL | Chinmaya Mission of Sri Lanka |
| Consortium of Mine Action Organizations | |
| COOPI | Cooperazione Internazionale (International Cooperation) |
| COWL | COWL |
| Cricket Aid | |
| CRS | Catholic Relief Services |
| CRWRC | Christian Reformed World Relief Committee |
| CTF | Community Trust Fund |
| DCA | Dan Church Aid |
| DCS | Developmental Cooperation Solidarity |
| DRC IFRC | Danish Red Cross |
| EFSL | Emace Foundation of Sri Lanka |
| EHED | Eastern Human Economic Development Centre/Caritas |
| EP | European Perspective |
| FAO | Food and Agriculture Organization |
| FCE | Foundation for Co-Existence |
| FHB | Family Health Bureau |
| FHP | Family Help Protection |

| | |
|---|---|
| FinRC IFRC | Finnish Red Cross |
| FORUT | FORUT International |
| FRC IFRC | French Red Cross |
| Free the Children | |
| GA | Global Action |
| GFA | Gospel for Asia |
| GMSL | Green Movement of Sri Lanka |
| GOAL | GOAL |
| GRC IFRC | German Red Cross |
| GSF | Gemi Sewa Foundation |
| GVC | Gruppo di Volontariato Civile – Group of Civil Volunteers |
| HBF | Humanitarian Brotherhood Foundation |
| HELP | Help from Germany |
| HELPAGE | Helpage Sri Lanka |
| HI | Handicap International |
| Hidaya Foundation | |
| HSL | Helvetas-Sri Lanka |
| Humanitarian Medical Assistance | |
| ICES | International Centre for Ethnic Studies |
| ICRC | International Committee of the Red Cross |
| IFRC | International Federation of Red Cross and Red Crescent Societies |
| IHR | Institute of Human Rights Sri Lanka |
| IIRO | International Islamic Relief Organization |
| IMC | International Medical Corps |
| Impact Foundation Sri Lanka | |
| IOM | International Organization for Migration |
| Irish-Sri Lanka Trust Fund | |
| ITDG So | Intermediate Technology Group-South Asia |
| ITI | ITI |
| JCCP | Japan Centre for Conflict Prevention |
| JDC | American Jewish Distribution Committee |
| JIA | Johanniter International Assistance |
| JICA | Japan International Cooperation Agency Sri Lanka |
| JRS | Jesuit Refugee Services |
| KAA | Kurier Aid Austria (Guarantee) Ltd |
| LEADS | Lanka Evangelical Alliance Development Service |
| LFI | Loving Friends International |
| MCSWO | Mother and Child Social Welfare Foundation |
| MDM | Medecins Du Monde |
| Medair | Medair |
| MERCY Malaysia | MERCY Malaysia Medical Relief Society |
| MercyCorps | Mercy Corps |
| MERLIN | MERLIN |
| MFCD | Muslim Foundation for Culture and Development Sri Lanka |
| MI | Malteser International |
| MOVIMONDO | Movimondo |
| MR | Mission Rainbow |
| MSF-S | Medecins Sans Frontieres Spain |
| MTI | Medical Team International |
| NCC-SL | National Christian Council of Sri Lanka |
| NDF | Norwegian Development Fund |
| Neelan Thiruchelvam Trust. | |
| NFDR | Northwest Firefighter Disaster Recovery Team |
| NMTI | Northwest Medical Teams International |
| NRC | Norwegian Refugee Council |
| OfERR | Organisation for Eelam Refugees Rehabilitation |
| OVS | OVERSEAS |
| OXFAM A | Oxfam Australia |
| OXFAM GB | Oxfam Great Britain |
| PHO | Polish Humanitarian Organisation |
| Pottuvil and Arugam Bay Fisherman Association | |
| RDF | Rural Development Foundation |
| ReC | Ricerca e Cooperazione |
| REIDO | Rural Industries and Enterprises Development Organisation |
| RI | Relief International |
| RI SOL | Relief International – Schools Online |
| RR | Room to Read |
| SAH | Schweizer Arbeiterhilfswerk (SOLIDAR Consortium) |
| Sarvodaya Shramadhana Movement | |
| SCF | Shilpa Children's Trust |
| SCI | Service Civil International |
| SCiSL | Save the Children in Sri Lanka |
| SDF | Saviya Development Foundation |
| SEDEC | Social Economic Development Centre |
| SEEDS | Sarvodaya Economic Enterprise Development Services |
| SFA | Stromme Foundation – Asia |
| SFL | Shelter for Life |
| Shadow | Shadow |
| SHF | Solid House Foundation |
| SLCCCA | Sri Lanka Churches Child Care Association |
| SLF | Sewa Lanka Foundation |
| SLRCS | Sri Lanka Red Cross Society |
| Small Fisher's Federation | |
| SOLIDARITIES | SOLIDARITES – Aide humanitaire d'urgence |
| SOND | Social Organisation Networking for Development |
| SOS | SOS Children's Villages of Sri Lanka |
| SPIR | Samaritan's Purse International Relief |
| Sri Lanka Direct | |
| Sri Lankan Human Rights Commission Success Sri Lanka | |
| TDH | Terres Des Hommes-Lausanne |
| TRO | Tamils Rehabilitation Organisation |
| TRSL | The Rebuild Sri Lanka Trust |
| UMCOR | United Methodist Committee on Relief |
| UN OCHA | United Nations Office for the Coordination of Humanitarian Affairs |
| UNDP | United Nations Development Programme |
| UNFPA | United Nations Population Fund |
| UNHCR | United Nations High Commissioner for Refugees |
| UNICEF | United Nations International Children's Fund |
| UNOPS | United Nations Office for Project Services |
| UNV | United Nations Volunteers |
| USAID | United States Agency for International Development |
| VSO | Voluntary Service Overseas |
| WDC | Women's Development Centre |
| Wes-Link | Wes-Link |
| WFP | World Food Programme |
| WHO | World Health Organisation |
| WUSC | World University Service of Canada (WUSC) Sri Lanka |
| WVL | World Vision Lanka |
| ZOA | Zoa Refugee Care – Netherlands |

# Siting conflict and peace in post-tsunami Sri Lanka and Aceh, Indonesia

JENNIFER HYNDMAN

The article aims to analyse two policy narratives that were politicised in the context of post-tsunami response in Eastern Sri Lanka and Aceh, Indonesia. These areas had been affected by war for several decades before the tsunami hit. The first narrative of public safety saw government imposition of post-tsunami buffer zones, ostensibly as measures to protect those affected, but they also incited tensions in both locations. The second relates to post-tsunami aid distribution. Developing policies and implementing bodies for tsunami aid has proven highly contentious, but uniquely so in each location. Whereas the tsunami in Sri Lanka has been followed by renewed fighting and the end of the 2002 Ceasefire Agreement, the province of Aceh enjoys the fruits of the 2005 peace agreement that has created greater autonomy from central government and greater access to resources than ever before. Each conflict has historically and geopolitically distinct antecedents, resulting in very different post-disaster policy contexts and political challenges on the ground.

*Jennifer Hyndman, Department of Geography, NTNU*

The tsunami of 26 December 2004 led to extraordinary devastation and destruction for peoples of the Indian Ocean Basin region. The earthquake that spawned it killed an estimated 167,000 people in Aceh, the closest land mass to the epicentre on the Indonesian island of Sumatra. In Sri Lanka, more than 36,000 people died, and another 800,000 were displaced from their homes. Overseas aid was unprecedented. In total, more than USD 13 billion was pledged, with more than USD 5 billion coming from private individuals and companies. Yet, in comparison the destruction and loss of life from the immense earthquake that struck Pakistan soon after hardly registered among private donors of the global North less than one year later (Grundy-Warr & Sidaway 2006).

Several of the states affected by the 2004 tsunami were host to long-term conflicts before the disaster, including Aceh and Sri Lanka, but also Burma (officially known as Myanmar), Somalia, and India. This coincidence of humanitarian crises – conflict on the one hand, and the tsunami on the other – raises important questions: What effect did the 2004 tsunami in the Indian Ocean Basin have on the extant contexts of conflict and peace in Aceh and Sri Lanka? What policy narratives were mobilised in the aftermath of the tsunami in each location? Did these tsunami-related policies intersect with the extant conflicts in both environments? This article aims to provide some preliminary responses to these questions.

Policy matters; there is no such thing as a 'natural' disaster (Smith 2006). A widely-held view within disaster management is that human decisions and actions shape the outcome of particular emergencies. For example, in 2008, Burma (or Myanmar in official parlance) was drastically affected by Cyclone Nargis, which killed 130,000 people and displaced over two million more along the Irrawaddy Delta. This violent storm and the waves it generated became a full-blown humanitarian disaster because of government *policy*. Virtually no international aid organisations or staff were allowed into the area to assist those affected until almost one month after the cyclone hit. The idea of a policy-induced disaster is pertinent to all humanitarian crises, whether it is related to conflict, tsunamis, cyclones, earthquakes, or other calamities.

Adding to existing literature on the intersection of 'man-made' versus 'natural' disasters (Wisner et al. 2004; Kelman & Gaillard 2007), this article juxtaposes geographies of conflict, devastation, and response in Sri Lanka and Aceh, Indonesia, two places acutely affected by the 2004 tsunami. A number of geographers (Korf 2005; Lawson 2005; Marston 2005; Olds et al. 2005; Stokke 2005; Kleinfeld 2007; Le Billon & Waizenegger 2007; Gaillard et al. 2008) and other social scientists (e.g. Jeganathan 2005; Kennedy et al. 2008) have taken up different elements of the question probing the relation of the tsunami to the conflicts that preceded it.

This article seeks to juxtapose the cases of Sri Lanka and Aceh in the context of post-tsunami aid and reconstruction. Methodologically, however, the work on which it is based is not strictly comparative research. The contexts of Sri Lanka and Aceh, Indonesia are simply too distinct to forge such an approach. The fieldwork for this article was conducted in both countries after the tsunami. Interviews with tsunami- and conflict-affected people in Sri Lanka took place in February 2005 and 2006. Interviews were also conducted with UN personnel and the staff of international non-governmental organisations (INGOs). In Banda Aceh, I held 27 interviews with UN personnel, government representatives, and INGOs in June 2007, as well as three individual interviews and one focus group with former combatants at that time.[1] In 2008, a further 50 interviews were completed with a cross-section of the same groups by research assistant Arne Waizenegger. This paper highlights issues and policies common to both contexts, but also examines responses at a more generalised level compared to a single case study.

Perhaps the most important caveat framing this research is that geographically diverse processes produced the particular patterns and examples provided below. As Matthew Sparke (2005, xiv) reminds geographers, 'any assumption about geography either as a result of or as a basis or container of spatial relations for other social relations always risks fetishising a particular spatial arrangement and ignoring ongoing processes of spatial production, negotiation, and contestation'. To overstate any geographical similarities is to ignore the multi-scalar actors and relationships that generate distinctive histories of struggle as well as contemporary policies and responses to the devastation of the tsunami of 2004. Contextualising the spaces of disaster, then, is an important departure point.

The people of Aceh have long fought for autonomy against colonial rulers as well as the centralised Government of Indonesia. In Sri Lanka, the political aspirations of a separate breakaway Tamil state by the Liberation Tigers of Tamil Eelam (LTTE) is a post-independence war, stemming from the anti-Tamil pogroms of 1983 in Colombo. Calls for a separate state in Sri Lanka can be traced to the early 1970s, when constitutional reforms alienated Tamils in what was formerly Ceylon. Both conflicts were in their third consecutive decade when the tsunami hit.

Today, several years after the 2004 tsunami event, political and economic conditions in the two affected areas are both dramatically different and starkly similar. In Aceh, the Indonesian Government signed a peace agreement with the Free Aceh Movement (*Gerakan Aceh Merdeka* or GAM) in August 2005, just eight months after the tsunami; it was still operational at the end of 2008. In Sri Lanka, political developments since the tsunami have *intensified* the country's prolonged political crisis (Uyangoda 2005), leading to the abrogation of the 2002 Ceasefire Agreement in 2008 by the Sri Lankan Government in the wake of rebel attacks. The number of civilian deaths since the tsunami continues to climb in Sri Lanka, while Aceh enjoys a precarious but continuous period of calm.

Despite this crude political contrast of peace and war, both places continue to host political tensions and resentment in the wake of tsunami reconstruction. The landscapes of conflict and displacement that preceded the tsunami have not been and cannot be erased. Rather, a variegated, politicised layer of aid overlays the earlier disparities, histories, and geographies of displacement. In Sri Lanka:

> [s]truggles over inter-ethnic justice, neo-liberalism, economic distribution, the disempowerment of women, caste bigotry and such have shaped the [Sri] Lankan political landscape in significant ways over the last decades ... even the tsunami cannot wipe out the imprint of these fault lines. (Nesiah et al. 2005)

As Michael Renner (2006) notes in the case of Aceh, the *ulamas* (Islamic theologians) have blamed women for the tsunami, calling them 'female sinners' whose dress, for example, was not morally appropriate prior to the disaster. Shari'a law pre-dated the tsunami but the Shari'a police did not emerge until afterwards. A feminist analysis of these political transformations remains to be done. Distinct societal cleavages, political tensions, and economic disparities between Aceh and Sri Lanka pre-date the tsunami, so any juxtaposition must avoid direct comparison that might eclipse the nuanced differences and dynamics that have historically and geographically constituted each place (Action Aid 2006).

## Political geographies of displacement: pre-tsunami

The devastation of the tsunami overlays longstanding geopolitical tensions, struggles for resources, and political geographies of displacement (Hyndman & de Alwis 2004; Nah & Bunnell 2005). In what follows I briefly sketch the political backdrop to conflict in both Aceh and Sri Lanka, setting the stage for the 'tsunami politics' that ensued after the fateful waves of 26 December 2004 hit Aceh and Sri Lanka.

### *Indonesia*

Sentiments towards sovereignty and Islam in Aceh are more pronounced than in most other parts of Indonesia. When Dutch forces recaptured Indonesia in 1946, after Japan's defeat in World War II, Aceh remained the only free region of the archipelago (Waizenegger 2007). Both Acehnese people and politicians broadly supported the creation of the Indonesian Republic in principle, in large part because they believed it would be an Islamic state and Aceh would be granted special provincial status (Aspinall 2007). The creation of a secular and centralised Federal Republic of Indonesia in 1949 betrayed Aceh's loyalty and support, sowing seeds of rebellion.

In response to this dissidence, Indonesian President Sukarno granted Aceh separate provincial status in 1959, and gave it the status of a 'special region' (*Daerah Istimewa*) in 1961 to pacify the Acehnese. However, Sukarno's political nemesis, President Suharto, stoked discontent during his rule, beginning in the mid-1960s. Feelings of exploitation soared among the Acehnese in 1971 when huge oil and liquid natural gas deposits were discovered near Lhokseumawe and Lhoksukon in North Aceh. Exploration was followed by the construction of what was then a huge refinery in worldwide terms, financed as joint venture between the Indonesian state-owned Pertafina and ExxonMobil (Renner 2006).

Other companies seeking to run the refinery included Doral International, a conglomerate owned by the Acehnese businessman Teungku Hasan di Tiro (Waizenegger & Hyndman in press). Di Tiro is a descendant of the last sultan of Aceh, and a direct descendant of Teuku Cik Di Tiro, a famous Acehnese national hero during the Aceh War of 1873–1903, so his defeat in the oil deal was seen as a renewed expression of the unjust Javanese economic exploitation and domination of the province. The confiscation of villagers' land without compensation by the Government of Indonesia and the deployment of as many as 5000 troops to protect ExxonMobil from the emerging rebel force also fuelled Acehnese nationalism (Renner 2006).

The Free Aceh Movement (GAM) has been fighting for Aceh's independence from Indonesia since 1976. After a number of attacks by GAM in 1989–1990 – a period of extensive human rights violations on both sides – the Indonesian military launched a massive counterinsurgency operation in Aceh Province which continued on and off into 2005 (Nah & Bunnell 2005). Conflict, violence, and the sustained counter-insurgency had displaced more than 300,000 people since 1999, well before the tsunami hit. In December 2002, a cessation of hostilities agreement was signed, drastically reducing the number of killings (Renner 2006). Yet, by May 2003, in response to renewed Acehnese calls for a share of resource revenues, the Government of Indonesia declared a state of emergency in Aceh (Hedman 2005), and the region was highly militarised by the Indonesian government which installed more than 40,000 additional troops to forcibly relocate insurgents.

In October 2004, Susilo Bambang Yudhoyono became President of Indonesia, with a pledge to seek peace in Aceh. Then, in December 2004, the tsunami hit Aceh. Just eight months later, after five rounds of talks, a peace agreement, or memorandum of understanding (MoU), was signed in August 2005, securing withdrawal of troops, the disarmament of rebels, more energy revenues for the Aceh region, and greater autonomy for local government in Aceh. As the wording of the peace agreement specifically states: 'The parties are deeply convinced that only the peaceful settlement of the conflict will enable the rebuilding of Aceh after the tsunami disaster on 26 December 2004 to progress and succeed.' The newly elected president put it this way: 'The tsunami produced an overwhelming moral, political, economic, social imperative to end the conflict ... I was criticised by those who did not see any benefit from renewed talks with GAM. But I was more concerned about the judgment of history for missing this rare window of opportunity to resolve the conflict' (President Yudhoyono, opening speech, 'Building Permanent Peace in Aceh: One Year After The Helsinki Accord' conference held in Jakarta, 14 August 2006). The principles of peace outlined in 2005 were passed into law governing Aceh, in diluted form due to resistance by hardliners, in July 2006 (Renner 2006).

Glassman (2005) argues that the Indonesian state was partially to blame for the susceptibility of people to death and destruction in Aceh during the tsunami. Despite Aceh's natural wealth in resources, Javanese sub-imperialism and the US-backed rule of President Suharto led to Acehnese 'people leading marginal lives and surviving on marginal resources' (Glassman 2005, 166). This confluence of cold war geopolitics, in which the West provided considerable backing for the Indonesian military and Java-centric governance that ensured energy revenues flowed from Aceh to Jakarta, generated a skewed political economy in the region. Such political currents continued even in the wake of the tsunami when, as Glassman reports, GAM honoured the ceasefire while the Indonesian military continued attacks on GAM strongholds.

## *Sri Lanka*

In Sri Lanka, a war characterised as one of violent competing nationalisms between the secessionist Liberation Tigers of Tamil Eelam (LTTE) and the Government of Sri Lanka's armed forces has been waged for more than 25 years (Jayawardena & de Alwis 1996, ix–xxii). Contemporary Sri Lanka is an expression of a long history and geography of struggle well-documented by Sri Lankan scholars (Abeysekera & Gunasinghe 1987; Jeganathan & Ismail 1995; Tiruchelvam 1996). At the time of writing (2008), the death toll from the conflict exceeds 70,000, and internal displacement follows as a consequence of the dynamics of conflict. During the Ceasefire Agreement between the Government and the LTTE, signed in February 2002, deaths fell dramatically until 2006, when hostilities resumed despite the Agreement. In a moment of donor hope in 2003, a conference to finance the reconstruction of war-torn Sri Lanka met in Tokyo; USD 4.5 billion was pledged for the 'Regaining Sri Lanka' strategy, referred to as a 'cash for peace approach' (Jeyeraj 2005; Bastian 2007a).

After the 2004 tsunami in Sri Lanka, the World Bank assessed the cost of reconstruction and recovery at approximately USD 1.5 billion. How to allocate these funds emerged as one of the biggest hurdles between the government and the LTTE. A joint mechanism to distribute international aid to tsunami-affected areas was proposed at an international donor forum in May 2005 (Kleinfeld 2007). Called the Post-Tsunami Operational Management Structure, or P-TOMS, the joint mechanism was a memorandum of understanding (MoU) that set out terms for a working relationship between the Government of Sri Lanka and the LTTE. P-TOMS had representation not only from the Sri Lankan Government and the LTTE, but also from Muslim political parties. Muslim communities living along the devastated east coast were disproportionately hit by the tsunami, given their overrepresentation in that region.

Signed in June 2005, the P-TOMS MoU was also a firebrand for political acrimony, galvanising Sinhalese nationalist sentiment in opposition to it because of the legitimacy it potentially conferred on the militant Tamil nationalists, the LTTE. The Sinhala nationalist Janatha Vimukthi Peramuna (JVP) party, in a coalition with the sitting government, petitioned the Supreme Court in July 2005 on the grounds that P-TOMS violated the rights of Sri Lankan citizens and the territorial integrity of the state on the following grounds (Kleinfeld 2007, 179):

- The LTTE was a terrorist organisation and not a governmental entity that could participate in such an agreement.
- The committees described in P-TOMS were governmental in nature and could not legally do the work they were charged to do without constitutional changes.
- Donor funds were funds of the Republic and could not be controlled by an outside agency such as the World Bank (as was outlined in the MoU).
- Treatment of persons within the tsunami disaster zone (TDZ) discriminated against tsunami-affected persons outside the TDZ.

If P-TOMS had succeeded, it might have served as a blueprint for the constitutional change required for peace beyond a ceasefire in Sri Lanka; however, it did not. The Supreme Court largely agreed with the plaintiff, and the MOU was never adopted. Then Prime Minister Mahinda Rajapakse ran for President later in 2005 on an anti-P-TOMS platform and won by a slim margin (Bastian 2007b). With his presidency and its consolidated alliance with Sinhalese hardliners, such as the JVP, violence has increased notably in Sri Lanka.

By early 2008, the Government of Sri Lanka had ended the ceasefire agreement and the Nordic monitors of the Sri Lanka Monitoring Mission (SLMM) had left the country. Bus bombs and suicide attacks in public areas, such as those witnessed throughout much of the late 1990s, have resumed. Between December 2004 and June 2007, more than 6000 people were killed (Kleinfeld 2007).

## Political geographies of displacement: post-tsunami buffer zone policies

The tsunami generally worsened the humanitarian, social, and economic situation of people living in the province of Aceh and in Sri Lanka. The regions most affected – Western Aceh on the northern part of Sumatra island and the eastern coast of Sri Lanka[2] – both have long histories of exclusion, poverty and conflict. Yet, the infusion of aid and the patterns it took created very different dynamics in each location. In both Sri Lanka and Aceh, money to rebuild houses was plentiful, but coordinating housing reconstruction and securing the land on which to build them were more elusive (Centre for Policy Alternatives 2005; Renner 2006). In both countries 'buffer zone' policies that initially prohibited rebuilding homes near the sea generated another wave of displacement, politicising humanitarian aid (Action Aid 2006). In the following section, I discuss a geography of displacement engendered by this policy. The buffer zone policy displaced yet again those who had already lost their homes through the dislocation of conflict and then the tsunami. In the case of Sri Lanka, this fuelled resentment among minority ethnic groups, namely Muslims and Tamils. Some Sri Lankan commentators have argued that the policy was politicised land grabbing on the part of a precarious government in one of the most densely populated countries in the world.[3]

Aceh, in contrast, experienced a different kind of politicisation in relation to the buffer zones; there, however, much more resentment based on unequal treatment of 'tsunami-affected' versus 'war-affected' people emerged in the form of protest. Below, I will address each of these contexts in turn.

### *Fostering conflict, buffering blame in Sri Lanka*

> When the buffer zone made its entrance after the waves had left behind the destruction [in January 2005], it was known by another name, less popular – set back zone. It most certainly has lived up to that title. (Perera 2005, 16)

The 'set back' or buffer zones declared by the Government of Sri Lanka represented a state-sponsored dislocation for many people along the east coast who had already been affected by the dual disasters of war and huge waves. Before the tsunami, during periods of intense conflict many Muslim and Tamil communities had found sanctuary on these unoccupied shores, which were narrow strips of land between the sea and the more inland lagoons, called littoral, or *eluvankaral* in Tamil). The hinterland, or *paduvaankaral*, farther inland from the lagoons is occupied largely by Sinhalese, the majority ethnic group, who were resettled from other parts of the country through government colonisation schemes during the nationalist periods of the 1950s and 1970s (Jeyeraj 2005).

From the time when the Sri Lankan Government announced the buffer zones, or no-build areas, cries of ethnic discrimination could be heard. Muslim communities were over-represented as a proportion of the general population in the hardest hit parts of eastern Ampara District and therefore over-represented in terms of the death and destruction wreaked in that area. So when the government announced a 200 m no-build zone along the east coast in contrast to a narrower 100 m buffer zone along the south coast, where homes of the Sinhala majority are most numerous, Tamil and Muslim groups called the buffer zones unfair and a sign of ethnic prejudice (de Alwis 2005). The zones served to reproduce extant patterns of discrimination against minority Tamils and Muslims in the wake of the tsunami and fuelled mistrust (Hyndman 2007).

In both areas, the high population density and scarcity of land also made the setbacks highly contentious. The President's office announced that the government would identify lands closest to the affected villages and build a house for every affected house owner who lived within the 100 m buffer zone (Sambandan 2005). It was further announced that owners of homes that were located inside the zone 'will retain the ownership of his original land' and that the government 'will not in any way claim ownership to such property'; moreover, the owner would be 'entitled to appropriate the land [within the 100 m zone] as he wishes, except building on it'. In addition, the government was to extend patronage to planting coconut or any other suitable crop on the land (Sambandan 2005). Land within the buffer zone could still be used for business purposes – a particularly important resource for tourist operators in Sinhala-dominated beach areas of southern Sri Lanka (MONLAR & ANRHR 2005). Also, hotels were rebuilt less than 10 m from the high tide water mark. So, while Sri Lankan citizens could no longer reside along the south coast, tourists could.

The terms of rebuilding and compensation were also contentious. Owners of homes and squatters alike could have houses rebuilt with aid either from the government or from an approved non-governmental organisation, but tenants renting from owners received nothing except the right to rent again if and when owners rebuilt (Centre for Policy Alternatives 2005). The logic, or lack thereof, behind the decision showed early signs of strain.

As the Institute for Policy Studies (2005) in Colombo stated, if public safety was the prevailing aim, the buffer zones should have been equivalent for all areas. The specific environmental, social and physical characteristics of coastlines in different parts of the country ostensibly require responses tailored to those geographies. Instead, Sri Lanka's World Bank representative, Peter Harrold, noted that buffer zones have been the single greatest barrier to progress in housing reconstruction for those who lost their homes in the tsunami (cited in Dias 2006).

The lack of evidence used to rationalise the buffer zones is striking, a situation that politicised the policy and intensified ethno-national tensions. On 28 March 2005, an earthquake of similar magnitude and near the epicentre of the 2004 tsunami shook much of Sri Lanka. President Chandrika Kumaratunga stated that the mere occurrence of the earthquake, which did not create a tsunami, was a logical rationale for the buffer zone: 'the people now should realise that the government, bearing in mind all allegation leveled against it, has acted prudently with a vision and in a responsible manner' (*Associated Press* 2005). Meanwhile, an opposition MP and former minister, Ravi Karunanayake, retorted that in his riding of Crows Island 1500 National Housing Development Authority houses went underwater in the tsunami, despite the fact that they were built more than a 100 m from the shoreline, which was proof, he said, that the buffer zone would be ineffectual.

The buffer zones proved to be a political hot potato for President Chandrika Kumaratunga, whose term ended in November 2005, at which time her then Prime Minister, Mahinda Rajapakse, was elected President of Sri Lanka. Rajapakse quickly distanced himself from the Kumaratunga presidency, first by changing his predecessor's government tsunami response body from the Task Force to Rebuild the Nation (TAFREN) to the Reconstruction and Development Agency (RADA). Then, through RADA, he announced in February 2006 that the buffer zone 'set back standards' would be relaxed and that the setback standards of the Coastal Zone Management Plan of 1997 would be revived (*The Sunday Times* 2006).

While the Advisory Council of the Coast Conservation Department approved exemptions from the buffer zone policy even before Kumaratunga left office (Cassim 2005), Rajapakse appeared to have a scientific reason for invoking the Coastal Zone Management Plan. The plan had been implemented but not enforced by various Sri Lankan governments since the 1980s and was based on protecting the coast and its inhabitants through appropriate vegetation. The 1997 version allowed for setbacks between 35 m and 125 m from the permanent vegetation boundary, depending on coastal conditions. That the plan pre-dated the tsunami and appeared to the public to be based on environmental science depoliticised the buffer zone controversy.

## *Aceh's blueprint and buffers*

In Aceh, the size of buffer zones announced by the government was substantially greater at the outset: 'Within a two-kilometre area from the shore, we will avoid building houses, offices, markets and shopping centres' (Mawardi Nurdea, head of Aceh's urban planning and housing authority in February 2005 cited in Sukarsono 2005). Fishers were exempt from this policy and could live 400m from the sea, but protests were raised nonetheless. There was little consultation with those displaced by the tsunami who abhorred the top-down policy (ACHR 2006). By March 2005, arguments against the buffer zones by those affected were faltering and community-based consultations were demanded by donors and beneficiaries alike. The blanket policy of buffer zones was lifted, as was evident during a field visit in 2007 that I made to Lok Nga, *c.* 10 km west of Banda Aceh, on the hardest hit west coast. There, an entire village had been rebuilt not more than 400 m from the sea, around a large mosque that had largely survived the tsunami. While many of the Turkish Red Crescent constructed houses were vacant due to poor (seasonal) water supply at the time, they have been considered among the best reconstructed buildings in the wake of the destruction.

In both cases, the buffer zone proclamations initially represented the position of central governments which then evolved and became more geographically tailored and politically-sensitive. In both Aceh and Sri Lanka, criticism in the immediate aftermath of the tsunami also focused on international non-governmental organisations' neglect in consulting with their national and local counterparts in the immediate aftermath of the tsunami (Couldrey & Morris 2005). As mentioned above, in Aceh the top-down policy was contested early on. In Sri Lanka, although coordination of tsunami aid began in the President's office, it soon became clear that such centralisation of control was not politically popular. Provincial civil servants met with international NGO (INGO) staff until late in most nights in early 2005, in attempts to allocate projects in affected areas to different organisations; to make peace with warring NGOs who claimed similar areas to rebuild, and to avoid duplicating services or aid provision. In Aceh too, INGOs competed with one another for space to conduct their humanitarian projects (Renner 2006).

One major policy distinction between the two contexts in the delivery of post-tsunami aid – at least at the outset – was the more centralised planning in the form of a *blueprint* by the Government of Indonesia in contrast to the more decentralised planning and coordination that eventually took place at the national, provincial and district levels in Sri Lanka, with international non-governmental organisations taking major roles and staking claims to particular projects and places. The relative influence of these private, if non-profit, international organisations compared to the Sri Lankan governing authorities at various scales remains an important issue that has been addressed by Sunil Bastian (2007b) in his recent book.

The pursuit of private for-profit interests in humanitarian reconstruction must also be addressed. In both Sri Lanka and Aceh, the rationale for the buffer zones was public safety, but as soon as the policy was announced in Aceh, private sector developers and other political insiders began jockeying for a piece of the post-tsunami reconstruction pie outlined in the blueprint (Renner 2006), attempting to profit from what Naomi Klein (2005) has called 'disaster capitalism'. Klein visited Sri Lanka after the tsunami and voiced concern that tsunami funds would be used for private contractors to build new port facilities and an inland superhighway which were not necessarily related to the destruction caused by the tsunami.

Researchers need to prove or dispute these ideas before reproducing them further. In Aceh, critics have recently raised concerns about why a new superhighway was built (not rebuilt) along Aceh's sparsely populated west coast after the tsunami. Speculation that the discovery of oil reserves off Aceh's west coast requires such road infrastructure and that international donors may have known about such reserves even before the tsunami occurred have circulated, but again this requires further investigation to prove or disprove its veracity.

### *Rich cousin, poor cousin: tsunami aid politics and post-conflict integration*

Just as disaster is never simply 'natural' (Smith 2006), 'reconstruction is never just a physical task' (Renner 2006, 13). In Sri Lanka along the eastern coast, tsunami-affected people were also intensely affected by conflict, especially in terms of geographical displacement. However, in Aceh, those who were intensely affected by conflict largely did not overlap with those devastated by the tsunami (Waizenegger & Hyndman in press). In Sri Lanka, communities of people displaced from conflict that were not affected by the tsunami have been largely ignored, at least in terms of tsunami aid. One distinction between conflict-affected people in Sri Lanka and Aceh is that the latter group experienced a dramatic loss in terms of livelihoods after the 2005 peace agreement (MoU, 2005) that required rebel fighters to demobilise and provided very limited funds for their reintegration. However, the MoU (2005) does make provision for reintegration in section 3.2.3:

> GoI [Government of Indonesia] and the authorities of Aceh will take measures to assist persons who have participated in GAM activities to facilitate their reintegration into the civil society. These measures include economic facilitation to former combatants, pardoned political prisoners and affected civilians. A Reintegration Fund under the administration of the authorities of Aceh will be established.

The MoU continues to state that all former combatants will receive an allocation of farmland, employment, or adequate social security. However, the number of combatants listed in the MoU (3000 military troops) is widely considered to be an undercount (Renner 2006) Hence the funds for reintegration are paltry. While some senior GAM leaders did receive a 'peace dividend' from the 2005 agreement in the form of jobs with the government agency for reconstruction (BRR, or Badan Rehabilitasi dan Rekonstruksi), the benefits were bestowed among a small, elite minority (Waizenegger & Hyndman in press).

In Aceh, deep-seated resentment has emerged based on the uneven distribution of funds between GAM commanders versus foot soldiers and on the dramatically different levels of aid available to tsunami survivors as compared to demobilised soldiers: 'The tsunami victims were the lucky ones ... At least they got help' (Ida Wati, former GAM rebel, cited in Honorine 2007); 'while almost USD 8 billion is committed to dealing with the aftermath of the tsunami, only USD 200 million is pledged for reintegration efforts' (Hollenbeck 2007).

A great number of GAM are frustrated, including combatants, non-combatants, and prisoners given amnesty:

> We also find a lot of dissatisfaction among GAM themselves. Their job is to fight, so now they are unemployed. They face problems feeding their families. There is a reaction to the fancy cars and 'Indonesian' behaviour that high ranking GAM have shown. GAM commanders got reintegration funds, but foot soldiers got nothing. There is a lot of resentment. (interview with a government representative in Banda Aceh, June 2007).

Reintegration is a highly political matter, with the potential for strong politicisation. The inadequacies of reintegration to date are explored in more depth by other authors (Waizenegger & Hyndman in press), but this pressing and current problem in Aceh distinguishes it from Sri Lanka, where the cessation of hostilities, let alone peace or reintegration, remains elusive.

## Without conclusion

Having briefly traced the unique antecedents to conflict and policy contexts in Aceh and Sri Lanka, this article has illustrated and analysed the broad contours of post-tsunami political geographies. In Aceh, evidence that the tsunami has accelerated and facilitated peace has emerged. In Sri Lanka, the history, terms and dynamics of conflict since the 2004 disaster are very different, but they have clearly intensified the country's prolonged political crisis. Policies ostensibly aimed at improving public safety in the event of another tsunami instead created antagonism. Government buffer zone policies to restrict rebuilding in designated setback areas along the coast proved highly contentious in both environments, but their outcomes were very different. In Sri Lanka, the buffer zones arguably fuelled ethno-national tensions if not outright conflict, whereas protest of this top-down policy by government (and the donor community) was met with a new, more consultative approach to post-tsunami reconstruction.

Likewise, various policy mechanisms, or blueprints, for *distributing* post-tsunami aid also proved to have significance beyond the mere logistics of reconstruction. In Aceh the peace agreement specified that some government jobs at the reconstruction agency (BRR) would be made available to ex-GAM members as part of a peace dividend. The blueprint created both the BRR agency for reconstruction and the BRA (Badan Reintegrasi Aceh) agency for reintegration, with formal (if contested) mandates and timelines for operation. In Sri Lanka, no joint mechanism was ever agreed upon. The Post-tsunami Operational Management Structure (P-TOMS) MoU was a promising political compromise while it lasted. Representation by Muslims, the LTTE rebels, and the Sri Lankan Government made it both an explosive but potentially radical intervention to enable humanitarian aid to be delivered. The political geographies of post-tsunami Aceh and Sri Lanka are thus, in one sense, a study in contrasts.

The political landscapes of Aceh and Sri Lanka today are strictly incomparable, yet their common histories of conflict and the tsunami of 2004 warrant further probing. In December 2006, Irwandi Yusuf, a former rebel leader with GAM, won the election for Governor of Aceh Province. Irwandi, who surprised most observers by being elected at all (he won 38% of the vote), was in jail for treason when the Indian Ocean tsunami crashed on Aceh in 2004. Unlike most of his fellow prisoners, he was able to swim to safety, escaping his cell only after the waves hit. Provincial elections in which newly formed parties will run are to be held in Aceh in 2009. Nonetheless, resentment is brewing among ex-GAM foot soldiers and those who lost their livelihoods with the cessation of hostilities. Disparities between the conflict-affected and the tsunami-affected also fuel the tension (Waizenegger & Hyndman in press).

All this is a far cry from Sri Lanka where a ceasefire agreement that had been in place for more than five years was cancelled in 2008, after two members of parliament (one a cabinet minister) were murdered and many more civilians killed in various bomb blasts in public spaces in the first half of the same year (Kleinfeld 2007). Prospects for a ceasefire thus looked increasingly slim, despite the hopes of donors who in 2003 pledged billions of dollars for post-conflict reconstruction. In 2009, a sustained Sri Lankan Government offensive against the LTTE continued. The chances of a ceasefire in this climate are slim to none.

More research needs to be done to ascertain the subtle and contested dynamics of sovereignty and governance *at multiple scales* in the post-tsunami relief and reconstruction period. The processes and policies employed by international aid organisations, national actors, and those affected by the dual disasters cannot be forced into a strictly comparative framework. The imperial designs and modalities of well-intentioned non-governmental organisations, with headquarters in Europe, Australia, Japan, and North America, may have been begrudgingly accepted by the tsunami-affected countries as the price to be paid for 'humanitarian assistance', but more research analysing these aspects of governance in the wake of so much aid is still needed. Likewise, 'disaster capitalism' (Klein 2005) in Aceh, Sri Lanka, and beyond has yet to be demonstrated. What is clear is that disaster aid has no singular effect on conflict. It may expand the political space to negotiate peace and at the same time breed discontent among those required to give up arms, as in the case of Aceh, or it may catalyse tensions among ethno-national groups in a war of competing nationalisms, as in Sri Lanka – a case of 'both/and'.

## Notes

1 Interviews in Aceh were interpreted with assistance from Arno Waizenegger. The project, funded by the Social Sciences and Humanities Research Council of Canada (SSHRC), aimed to trace the different responses by aid agencies to the tsunami.

2 The entire length of the Sri Lankan coastline was affected. The World Bank estimated that 40% of damage was along the east coast, 30% along the southern coast, 20% in the north (another region seriously affected by conflict), and 10% along the west coast.

3 I thank Sunil Bastian for his astute observation on this point.

*Acknowledgements.* – The author would like to thank the Social Sciences and Humanities Research Council of Canada (SSHRC) for funding the research. Thanks are also due to Piers Blaikie for his insights and suggestions for improving this manuscript. Thanks are also due to Ragnhild Lund for the encouragement to write this paper, and to Sunil Bastian, Eva-Lotta Hedman and Mala de Alwis for their comments on earlier drafts. I owe much gratitude to Arno Waizenegger, who was a remarkable research assistant in Aceh in 2007 and 2008.

## References

Abeysekera, C. & Gunasinghe, N. 1987. *Facets of Ethnicity in Sri Lanka*. Social Scientists' Association, Colombo.

ACHR (Asian Coalition for Housing Rights). 2006. *Tsunami Update.* www.achr.net/000ACHRTsunami/Download%20TS/Tsunami%20 Update%20June%20full%2006.pdf (accessed December 2008).

Action Aid. 2006. *Tsunami Response: A Human Rights Assessment*. Johannesburg. www.actionaid.org/docs/tsunami_human_rights.pdf (accessed 19 October 2008).

Aspinall, E. 2007. Guerillas in power. *Inside Indonesia*. http://insideindonesia.org/content/view/616/47/ (accessed December 2008)

*Associated Press.* 2005. Sri Lankan President pushes for controversial buffer zone after tsunami scare. 29 March 2005. Associated Press, Colombo.

Bastian, S. 2007a. *Globalisation, Foreign Aid and Conflict – The Case of Sri Lanka*. International Centre for Ethnic Studies Occasional Paper. International Centre for Ethnic Studies, Colombo.

Bastian, S. 2007b. *The Politics of Foreign Aid in Sri Lanka*. Colombo, International Centre for Ethnic Studies.

Cassim, N. 2005. Govt. gets flexible on buffer zone rule. *Daily Mirror* (Sri Lanka) 10 October 2005.

Centre for Policy Alternatives. 2005. *Landlessness and Land Rights in Post-Tsunami Sri Lanka*. Commissioned by International Federation of the Red Cross, Colombo.

Couldrey, M. & Morris, T. 2005. UN assesses tsunami response. *Forced Migration Review* Special Issue July 2005, 6–9.

de Alwis, K.A. 2005. The 100-metre rule – What's the logic? *The Island* 25 May 2005.

Dias, S. 2006. Key donors divided over post tsunami work. *Daily Mirror* (Sri Lanka) 7 February 2006.

Gaillard, J.-C., Clavé, E. & Kelman, I. 2008. Wave of peace? Tsunami disaster diplomacy in Aceh, Indonesia. *Geoforum* 39:1, 511–526.

Glassman, J. 2005. Editorial: Tsunamis and other forces of destruction. *Environment and Planning D: Society and Space* 23:2, 164–170.

Grundy-Warr, C. & Sidaway, J. 2006. Political geographies of silence and erasure. *Political Geography* 25:5, 479–481.

Hedman, E.-L. 2005. Introduction. *Forced Migration Review* Special Issue July 2005, 4–5.

Hollenbeck, L. 2007. *...and Then There Were Elections. Now What?*. USINDO, Washington, DC.

Honorine, S. 2007. Tsunami's tide of goodwill leaves rebels in the cold. *South China Morning Post* 10 September 2007.

Hyndman, J. 2007. Fear and conflict in post-tsunami Sri Lanka. *Annals of the Association of American Geographers* 97:2, 361–372.

Hyndman, J. & de Alwis, M. 2004. Bodies, shrines, and roads: Violence, (im)mobility, and displacement in Sri Lanka. Gender. *Place and Culture* 11:4, 535–357.

Institute for Policy Studies. 2005. *Sri Lanka: State of the Economy 2005*. Institute for Policy Studies, Washington DC.

Jayawardena, K. & de Alwis, M. 1996. *Embodied Violence: Comunalising Women's Sexuality in South Asia*. Kali for Women, New Delhi.

Jeganathan, P. 2005. SouthPaw One. *Lines* 3:4. www.lines-magazine.org/Art_Feb05/Pradeep.htm (accessed 12 October 2008).

Jeganathan, P. & Ismail, Q. 1995. *Unmaking the Nation: The Politics of Identity and History in Modern Sri Lanka*. Social Scientists' Association, Colombo.

Jeyeraj, D.B.S. 2005. The state that failed its people. *The Sunday Leader* 6 February 2005, 17.

Kelman, I. & Gaillard, J.-C. 2007. Disaster diplomacy in Aceh. *Humanitarian Exchange: Practice and Policy Notes* 37, 37–39.

Kennedy, J., Ashmore, J., Babister, E. & Kelman, I. 2008. The meaning of 'build back better': Evidence from post-tsunami Aceh and Sri Lanka. *Journal of Contingencies and Crisis Management* 16:1, 24–36.

Klein, N. 2005. The rise of disaster capitalism. *The Nation* 2 May 2005.

Kleinfeld, M. 2007. Misreading the post-tsunami political landscape in Sri Lanka: The myth of humanitarian space. *Space and Polity* 11:2, 169–184.

Korf, B. 2005. Sri Lanka: The tsunami after the tsunami. *International Development Planning Review* 27:3, i–vii.

Lawson, V. 2005. Natural disaster or space of vulnerability. *AAG Newsletter* 40:3, 3–4.

Le Billon, P. & Waizenegger, A. 2007. Peace in the wake of disaster? Secessionist conflicts and the 2004 Indian Ocean tsunami. *Transactions of the Institute of British Geographers* 32:3, 411–427.

Marston, R. 2005. Geography and the Indian Ocean tsunami. *AAG Newsletter* 40:2, 1, 4.

MoU (Memorandum of Understanding between the Government of the Republic of Indonesia and the Free Aceh Movement). 2005. http://old.thejakartapost.com/RI_GAM_MOU.pdf (accessed December 2008).

MONLAR & ANRHR (Movement for National Land & Agricultural Reform and Alliance for Protection of National Resources and Human Rights). 2005. Tsunami update II. *Lines* 3. www.linesmagazine.org/tsunami/MonlarJan22.htm (accessed 12 October 2008).

Nah, A.M. & Bunnell, T. 2005. Ripples of hope: Acehnese refugees in post-tsunami Malaysia. *Singapore Journal of Tropical Geography* 26:2, 249–256.

Nesiah, V., Nanthikesan, S. & Kadirgamar, A. 2005. Post-tsunami reconstruction – new challenges, new directions. *Lines* 3:4. http://www.lines-magazine.org/tsunami/linestsunamivision.htm (accessed 12 October 2008).

Olds, K., Sidaway, J.D. & Sparke, M. 2005. Editorial: White death. *Environment and Planning D: Society and Space* 23:4, 475–479.

Perera, A. 2005. The buffer zone fiasco. *The Sunday Leader* 25 December 2005, 16.

Renner, M. 2006. Unexpected promise: Disaster creates an opening for peace in a conflict-riven land. *World Watch* 19, 10–16, November/December 2006. http://www.worldwatch.org/node/4663 (accessed 27 February 2008).

Sambandan, V.S. 2005. Tsunami and rehabilitation: The rebuilding phase. *Frontline: India's National Magazine from the publishers of The Hindu* 22:4, 12–25 February 2005.

Smith, N. 2006. There is no such thing as 'natural' disaster. Published on 11 June 2006 in the SSRC (Socal Science Research Council) series 'Understanding Katrina: Perspectives from the Social Sciences'. http://understandingkatrina.ssrc.org/Smith/ (accessed December 2008).

Sparke, M. 2005. *In the Space of Theory: Postfoundational Geographies of the Nation-State.* University of Minnesota Press, Minneapolis. http://understandingkatrina.ssrc.org/Smith/ (accessed 29 February 2008).

Spencer, J. (ed.) 1990. *Sri Lanka: History and the Roots of Conflict*. Routledge, London and New York.

Stokke, K. 2005. After the tsunami: A missed opportunity for peace in Sri Lanka? *NIASnytt* No. 2, June, pp. 12–13. Copenhagen.

Sukarsono, A. 2005. Tsunami-hit Indonesia to get buffer zone. Reuters Foundation, posted on Relief Web. www.reliefweb.int/rw/rwb.nsf/db900SID/DDAD-69DUXH?OpenDocument (accessed 10 February 2008).

*The Sunday Times.* 2006. Coast conservation buffer zone limits relaxed. RADA advertisement in *The Sunday Times* 5 February 2006, 6.

Tiruchelvam, N. 1996. Sri Lanka's ethnic conflict and preventive action: The role of NGOs. Rotberg, R. (ed.) *Vigilance and Vengeance: NGOs Preventing Ethnic Conflict in Divided Societies*, 147–164. Brookings Institution, Washington DC.

Uyangoda, J. 2005. Ethnic conflict, the Sri Lankan state and the tsunami. *Forced Migration Review* 24. www.fmreview.org/pdf/uyangoda.pdf (accessed 12 June 2007).

Waizenegger, A. 2007. *Secessionist Conflicts and the 2004 Indian Ocean Tsunami.* Canada Asia Commentary Vol. 43. Asia Pacific Foundation of Canada (APFC), Vancouver.

Waizenegger, A. & Hyndman, J. In press. Two solitudes: Post-tsunami and post-conflict Aceh. *Disasters.*

Wisner, B., Blaikie, P., Cannon, T. & Davis, I. 2004. *At Risk: Natural Hazards, People's Vulnerability and Disasters.* Routledge, London.

# Index

Page numbers in *Italics* represent tables.
Page numbers in **Bold** represent figures.